FUNCTIONAL DESIGN FOR 3D PRINTING

DESIGNING PRINTED THINGS FOR EVERYDAY USE

THIRD REVISED EDITION

CLIFFORD SMYTH

~∞~DEDICATED TO THE INNOVATORS, MAKERS, AND DREAMERS ~∞~

WITHOUT WHICH THE WORLD WOULD BE A VERY DULL PLACE INDEED.

INTRODUCTION:

FUNCTIONAL DESIGN FOR 3D PRINTING is written as a practical design guide for implementing useful, practical objects using fused filament desktop manufacturing. This is not an in-depth engineering study of fused filament manufacturing, but rather a practical, hands-on treatment of the many engineering and design details specific to the 3D printing of useful objects.

The book is intended to achieve these aims while being accessible to the initiated layperson or 3D designer, without requiring an extensive background in manufacturing, engineering, or mechanics.

It is not a book *specifically* written for beginners in the field of 3D design, though it is hoped that the inquisitive novice may find it useful as a reference while resolving or avoiding issues in their projects.

This book focuses primarily on the popular FFM, (*fused filament manufacturing, also called FFF or FDM*™) process, where a plastic filament is fused into

an object layer by layer. Its contents may also have applications in other printing modalities, but these are not specifically addressed.

The field of 3D printing is a fast developing one, and surely before many of you read this some of the particulars may be dated. It is for this reason that much of the focus is on general principles of FFM manufacturing, rather than printer specific settings, data, or formula.

NOTE TO PROFESSIONAL USERS

Many references in this book are specific to open source printing software. Production stacks for professional printers can vary considerably and sometimes entire sections of the stack, such as the program that converts the 3D model into machine specific instructions (the slicer), are completely hidden from the designer and operator. If this applies to you, feel free to ignore these sections; but being aware of the behind the scenes machinations of your software can also help you to design more printable parts in edge-cases where even the best software may encounter difficulty.

Many commercial machines can produce more precise or reliable results than the open source platforms that are the primary focus of this book. I anticipate that as time progresses, this gap will narrow as consumer / prosumer printers converge on the industrial printing space.

In this book, I assume that designs may be printed by a wide audience of relatively inexperienced printer users, using a variety of open source software. Of course the results will vary, but the idea is to create models that will, to the extent possible, tolerate these variations.

TABLE OF CONTENTS

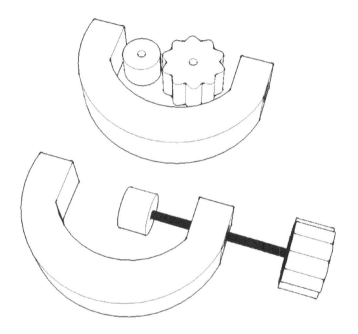

CHAPTER I: THE MEDIUM

Understanding the details, strengths, and weaknesses of FFM manufacturing: A brief summary of the medium, from a designer's perspective.

In addressing the unique considerations for modeling objects to be manufactured using 3D printing, I will be focusing on the FFM (*fused filament manufacturing*) process. These considerations may also apply to other printing processes in varying degrees. It is worth noting that the FFM process is identical in all respects except name to the Stratasys patented FDM™ process. FFM is also the same as FFF (Fused Filament Fabrication) which is a term that seems to be gaining traction.

As in any type of industrial design, modeling objects for 3D printing requires careful consideration of the manufacturing process. Although 3D printing is highly automated and requires minimal interaction, there are many important factors to be addressed in the design phase if the manufactured object is to print reliably and to perform anything other than an ornamental purpose.

DESIGNING FOR USABILITY

Apart from addressing the basic form of an object, the designer must also ensure that the model expresses the required strength and utility to perform its intended purpose. Notably, this functionality must be preserved through the printing process, which is the focus of this book.

In the case of purely ornamental objects where only cursory thought is given to structural strength and use, the designer's focus can remain on artistic expression and assuring that the model can be reliably printed as designed.

For utility oriented items the design process may be much different. Any desired characteristics other than the pure form of the object will need to be identified in the design process and incorporated by the designer as an engineered feature of the model.

Common factors to be considered include the magnitude and the direction of applied forces, the desired weight and size, requirements for flexibility and the amount of material and time that will be used in the printing process. Functional elements such as latches, hinges, or fasteners may also need to be designed into the model.

An important engineering concern to keep in mind as a designer is the anisotropic (non-uniform) strength characteristics of FFM printed objects. Due to the layer-wise construction process of fused filament modeling, printed parts will have relatively higher tensile strength (but shear more easily) along the X-Y (horizontal as printed) axis. Tensile strength in the Z (vertical as printed) axis can be comparatively low, but Z-axis compressive strength is typically good.

Often, models that could possibly be printed as a single object must be broken down into multiple components. This might be required in order to meet necessary strength, printability (print reliability), or material use criteria. This process of breaking a model down into sub-assemblies is known as factoring (or re-factoring) the design.

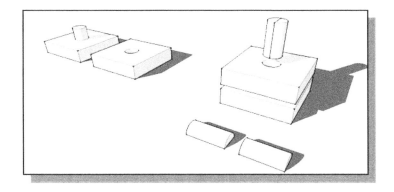

Design options for stacking blocks. Option one is to print one part with an integral pin (pictured to the left). While this seems like it might be a good idea, due to the anisotropic nature of FFM structures it will be weak and at risk for poor fitting. A better option would be to print two of the blocks with holes and the split pin pictured to align them (left). This leverages the strengths of the printed material and increases the probability of a successful assembly. (Illustration by author)

THE PALETTE: MATERIALS FOR FFM MANUFACTURING

The choice of material is a primary consideration when designing functional objects. Usually, the intended use of the object being designed will dictate which material is best used in its construction.

Sometimes, especially with less demanding applications and careful design, this requirement can be minimized so that the object will provide good service when printed in many different plastics.

Many thermoplastics are suitable for the FFM/FFF process, with more being optimized for use every month. Of these, ABS and PLA are the most widely used, mainly because they offer good structural characteristics without requiring special equipment, handling, or drying. Polycarbonate, TPU, PET, and Nylon are other common choices - each type of plastic possessing its own unique features and drawbacks.

Although plastics vary in their resistance to environmental factors, additives, and colorants contribute to these characteristics at least as much as the base material. For example, PLA is generally considered to be UV resistant, but some users report significant weathering over a short time span with some filament formulations. Likewise, I have used ABS filaments in applications lasting for years in the tropical sun, while other ABS formulations disintegrated in a matter of months. In general, darker colors resist weathering better, but "your mileage may vary" definitely applies here.

A catalog of common 3d Printing materials and their properties can be found in Appendix I, on page 168.

THE CANVAS: UNIQUE CHARACTERISTICS OF FFM STRUCTURES

Most FFM printed structures consist of a stack of layers. The printing process starts with a thin layer of plastic, forming the base of the part, which is extruded onto the print surface or bed. This is done using a small nozzle that extrudes the plastic onto the surface in a line, similar to drawing with a pen.

First, the edges or perimeters of the base are drawn. This will consist of one or more lines drawn at the outer (and inner, if applicable) "perimeters" or edges of the base layer. Then, to form a solid base, the area inside the perimeter(s) will be "colored in" or infilled with plastic, just like shading an area with a marker or pen. When the base layer is finished, it forms a thin sheet or "slice" of the base of the model a fraction of a millimeter thick.

The next layer or "slice" of the model is then drawn on top of the first, so that the part gradually increases in height, a fraction of a millimeter at a time. In this way, the model is eventually fully developed as a series of sub-millimeter layers, each layer fused to the one below it.

As described above, a printed object would always be built as a solid form. If the infill or "coloring in" phase of each layer was skipped, it would result in a hollow object. One of the great advantages of 3D printing is that the layer by layer paradigm enables a complex internal structure to be specified instead of only a solid or hollow one, so let's examine that process.

If for each layer, a grid was drawn instead of "coloring in" between the perimeters, space inside the perimeter would be filled with square hollow cells from top to bottom. This is a common infill methodology for low density (<25%) infill percentages. Another common infill algorithm includes parallel (or almost) lines that alternate orientation with each layer, forming a fibrous internal structure reminiscent of the material found in the lightweight bones of birds. Yet another strategy is to draw a hexagonal grid, resulting in a honeycomb structure that is both very strong and lightweight.

Infill design and density has an effect on the strength of the model, as does skin (perimeter) thickness and the number of solid layers at the upper and lower facing surfaces of the print. In most cases, these parameters are specifications given to the printing software rather than being explicit in the design.

The type and density of the infill are provided by the designer as a specification rather than explicitly represented in the model, and the infill plan is automatically generated during the pre-print processing or "slicing" of the model. In some very critical cases, it may be desirable to actually design the infill structure, but this is rarely practical or necessary.

It is worth reiterating that in all of these examples, the infill structure is automatically generated and is not directly expressed in the design process in most cases.

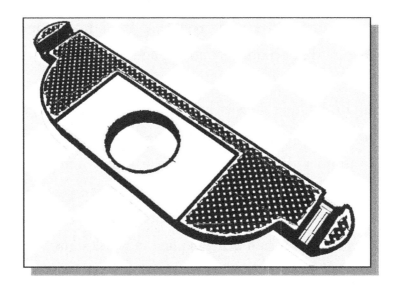

Slice view from a popular slicing software, CURA. Here the 50% crosshatch infill pattern can be clearly seen in the pictured layer (Illustration by author)

ANISOTROPY

One of the major structural factors affecting the strength and flexibility of FFM printed objects is the "grain" of the print. This characteristic arises from the printing process because fusion between layers of a print is usually weaker than fusion within a single layer.

Superficially similar to wood, 3D printed objects are anisotropic (not equal in all directions) in structure and strength. The relative weakness in interlayer fusion means that the layer-perpendicular Z-axis will have less tensile strength than the layer-wise X and Y axes. The amount of this variation in strength is primarily dependent upon the material being printed, the printer calibration, and the extrusion temperature. Unlike wood, the X and Y-axis strength orientation is determined by the design outline and infill orientation. This gives 3d printed objects 2 potential axes of good tensile performance and 1 weak one. Wood, on the other hand, has 2 weak axes and only one axis with high tensile strength.

While the tensile strength of the Z-axis is relatively poor, the compressive strength of the Z-axis is very good and can be enhanced through variations of internal structure, or in the density or pattern of the infill. *(also see the anisotropy lab on page 199)*

Generally, more infill will result in greater compressive, tensile and shear strength – up to a point. In most designs, infill above 60 percent starts to produce diminishing returns. Careful design can often do 'more with less' compared to relying strictly on high infill density for strength.

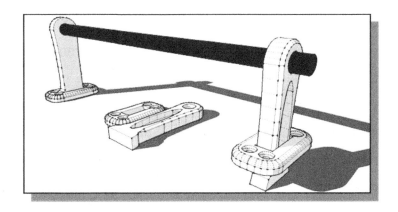

A towel rack. Note the parts are printed flat on the print bed and are assembled into their useful form. Grain orientation is optimized for strength. The beveled arm is held in place against the mounting surface by the mounting screws. (Illustration by author)

To get a feel for the effect of anisotropy or grain on part strength, it is useful to print test pieces and crush, tear or twist them to failure if you have access to a printer.

Observe the way the part fails.....usually, it will involve separation along the layers or "grain" of the print.

This "delamination" or interlayer failure will be one of the primary structural considerations as you design objects for practical use.

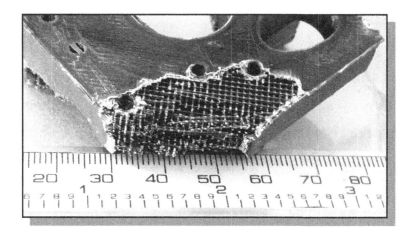

A broken 3D printed part. Note the tendency to break along the layers of the print and the internal structure, or 'infill'. (Photo by author)

SLICES OF STRUCTURE

Prior to printing, a model must be converted by pre-print processing software into a series of "slices", comprised of the model as a whole divided into horizontal layers. These slices are then expressed as detailed instructions for the printer to extrude each layer in turn, forming the finished piece one layer at a time.

Though the thickness of these slices is nominally a production issue rather than a design issue, slice thickness can also be a very influential engineering factor.

When designing thin or small objects, for example, it is sometimes desirable to design layer by layer, or slice by slice. In this method, the designer designs part or the entire model in sub-millimeter increments, specifying the exact geometry of some or all of each layer in the finished object. This gives the designer optimal control over the final structure of the printed piece, with the disadvantage of significantly increasing design time. Design by slice also requires the print to be sliced and printed at a very specific layer thickness - increasing the technical burden on the printer operator.

Design-by-slice can make scaling of the model difficult or impractical outside of a certain range. Scaling difficulty may actually be a design goal in some cases, where limiting a model to a specific size is desired.

Parts that must meet tight size specifications should ideally be designed for a layer thickness that is a factor of their height, as a fractional layer height multiple may reduce the dimensional accuracy of the finished part in the Z-axis. For example, if a part must be 20.5-mmin height, a layer height of 0.1mm, 0.125-mm or 0.25-mm might be specified by the designer, as a layer height of 0.2-mm would result in a part either 20.4-mm 20.6-mm in height.

Intricate models will generally print with more detail at a small slice or layer thickness. Gradual sloping or curved upper or lower surfaces will show aliasing errors more prominently with thicker layer heights, so thinner layers can be specified to help smooth out the errors, giving the finished print a smoother, more attractive appearance.

In addition to determining the texture or "resolution" of the Z-axis, layer thickness is also -the- dominant variable in printing time, all other things being equal. The same model sliced at 0.2-millimeters instead of 0.1-millimeters, for example, will print in approximately half of the time required for the thinner sliced model. Increased printing time implies an equivalent increase in the overall complexity of the printing process, which in turn increases the probability of a print failure.

Maximizing slice thickness where practical is often desirable for the potential improvement in print reliability alone.

To maximize the usable slice thickness for a given design, try to minimize the use of curves or slopes on the Z-axis, as they will show aliasing errors, or "steps" creating surface irregularities. Keeping surfaces close to vertical and precisely horizontal where practical will reduce print complexity, reduce printing times, and improve reliability.

Often, a part with complex surfaces can be reoriented to place the complexity in the X-Y plane, making the use of thin layers unnecessary to achieve smooth surfaces.

Slice or layer thickness can also influence some of the more dynamic physical characteristics of the printed piece. In general, thinner slices enable the printing of lighter pieces by incorporating thinner upper and lower surface shells. Thin layers can also be used for printing detailed membrane structures which can be useful as a structural web in ultralight assemblies, or more directly useful in features such as airfoils, gaskets, and foldable shapes.

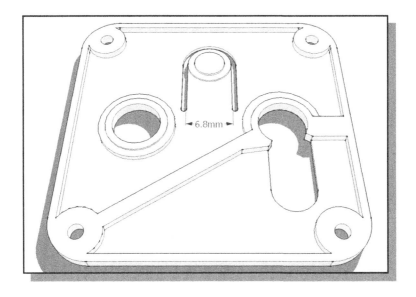

This pump gasket with integral flapper valve is designed slice-by-slice for printing at a 0.2-mm layer height. Functional membrane structures such as this can be designed for Nylon, TPE, or other flexible materials. (Illustration by author)

Close-up showing texture variations as a result of layer height. The model on the left was printed at 0.1mm, while the one on the right was printed at 0.25mm. Models that incorporate fewer curves or slopes near but not on the X-Y plane will be less sensitive to layer aliasing, facilitating faster, more reliable printing without sacrificing finish quality. (Photo by author)

THE PRINT NOZZLE... WHY IT CAN MATTER IN DESIGN

In FFM printing, plastic is extruded out of an extruder nozzle and deposited to make up the part, layer by layer. The width of this extrusion will depend partly on the nozzle width, and this can factor into design in several ways. For example, if a printed feature is to be two-millimeters wide and should be made solid, the required settings for the slicer will be different to accomplish this for a 0.5-millimeter nozzle than for a 0.4-millimeter one. It will also be impossible, for example, for a printer with a 0.5-millimeter nozzle to create hollow structures that are one-millimeter wide or thinner. These factors may need to be considered by the designer and any relevant notes passed along to the end user along with the model file.

SHELL AND INFILL SPECIFICATIONS

Many objects must meet specific strength or durability targets to be useful. The main factors affecting strength aside from the design of the model and the type of plastic used are the shell thickness and the amount of infill used when printing the part.

The types of plastic that will provide the desired characteristics, along with shell and infill specifications, should be provided by the designer as a readme file or as other documentation accompanying the design file.

As a starting point for shell specifications, 2-4 layers for top and bottom surfaces, and 1-4 perimeters or shells will usually give good results. A single shell or layer can be subject to weakness and permeability issues and is greatly improved by a second shell or layer. Using three or four layers and shells gives extra reinforcement to parts subjected to high stresses and adding shells (perimeters) helps to reinforce fastener holes and narrower parts. More than 4 layers or shells can be used, but lower quality filament or poor printer calibration can sometimes cause problems with excessive filling – potentially leading to print failure.

Infill, which is the structure printed in the otherwise empty space inside the shell, will be a factor in the strength, weight, printing time, and flexibility of the model. The plan for this structure is automatically generated by the slicing software, but can be altered parametrically in the slicer controls by the printer operator.

The primary structural design considerations for infill are orientation, geometry, and density. For multi-extruder printers, the choice of infill material may also be a consideration.

The percentage or density of infill determines the amount space inside the shell that will be filled by the infill plastic, so a 0.25 or 25% infill will fill 25% of the available space with extruded material. This is applied with varying precision depending on the software used, but generally, 100% infill creates a theoretically solid model. If it is desired to print at a very high infill density, it is usually best to specify around 90% as a maximum, because imperfect calibration or filament variation can cause print failures if the model becomes "overfilled" with plastic due to slight excess extrusion.

As a rule of thumb, infill gives a nominally linear improvement in part strength, with returns on strength versus material use and printing time diminishing at 45-70%, depending on part shape and loading direction.

As infill increases, flexibility may decrease - leading to brittleness as edges begin to fail in tension before bending, while less filled structures tend to be slightly more flexible.

Usually, the material or print time that could be used for very high infill percentages is better used by increasing part size, optimizing its geometry, or engineering its internal structure for optimal performance.

Empirically, infill over 60% is rarely advantageous, but may still prove useful when Z-axis tensile strength is critical or the part will be subject to severe compression or shear loads.

Since pure X-Y *tensile* strength does increase nearly linearly with infill up to around 90 percent, elements that will be heavily loaded in tension may be usefully infilled to very high percentages - but even in these cases, a redesign can usually give more strength than heavy infill for the material used.

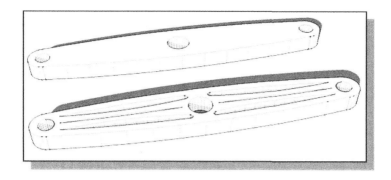

Two similar tensile elements. The lower one is much stronger in tension due to the high percentage of extrusion lines along the line of strain. It will also have the advantage of being able to flex more, aligning its elements with the precise vector of the applied force. Both examples use almost precisely the same amount of plastic to print. (Illustration by author)

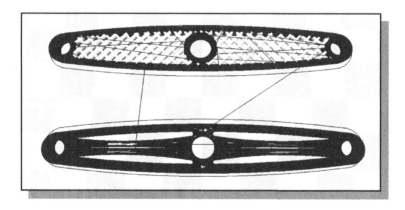

Slicer view of the two beams from the previous illustration. Notice how the lower beam has almost all of the extrusions running along the direction of strain and is nearly solid. (Illustration by author)

The primary objective in designing tensile elements should be the orientation and maximization of skin structures in the direction of strain. One strategy to accomplish this is to create multiple linear elements oriented along the direction of strain instead of a monolithic form. This usually ends up using a similar amount of material and printing time but aligns the extrusion fibers more optimally for strength and flexibility. This method can be especially useful with composite materials that contain reinforcing fibers, and can be used to develop very high performance structures

Five infill patterns – rectilinear, line and Triangle (top) with honeycomb and cubic (bottom), left to right. All were printed at 25% infill and sliced using Slic3r open source slicing software. (Photo by author)

COMMON INFILL GEOMETRIES

Some slicing software allows the operator the choice of infill form and orientation as well as density. The type of infill selected will significantly influence the speed of production and structural characteristics of the finished piece, so this can be exploited by the designer in some cases to good effect. Though there can be many possible infill options, the most common infill structures are rectilinear, line (shortest path), and honeycomb.

RECTILINEAR:

Where good overall compressive and tensile strength is desired with moderate flexibility and good printing speed, rectilinear infill is the go-to solution. It provides a predictable, uniform fill in a square interleaved pattern down to roughly 25% or even lower in almost all printers with good top layer finishing. Some printers will be able to make a smooth top layer with a much lower infill setting, depending on the size and shape of the print. Rectilinear infill works well with flexible filaments.

LINE:

Where speed of printing is the primary consideration, line infill is the ideal choice. It produces a somewhat irregular pattern of roughly square interleaved infill, but because the intersections cannot be relied upon to align throughout the Z-axis in complex parts, compressive strength is somewhat less than other more geometrically repetitive solutions. Torsional rigidity tends to be lower on narrow parts and this quality can be exploited to advantage in flexible structures. The principal advantages of line infill are its light weight, its printing speed, and that it offers a more harmonically complex structure that can be useful in creating parts that have a higher resistance to resonant vibration. Line infill is not well suited to printing with flexible filaments due to the lack of support under the individual extrusions.

CUBIC:

For overall rigidity, strength, and durability, cubic infill is an excellent solution. Very lightweight for the strength provided, it resists twisting, flexing, and compression in all axes. It is perhaps the most isotropic of infill types from a structural perspective.

Relatively large cavity sizes may require slightly higher density settings or upper surfac thickness to provide smooth coverage with some filaments and settings. Mixed results are probable with flexible filaments due to the overhanging geometry. Cooling may mitigate this effect.

HONEYCOMB:

For compressive strength in the Z-axis and skin rigidity for beam type parts, honeycomb infill can be a very good solution. The honeycomb structure is the same geometry used in many aerospace applications and it provides outstanding strength and rigidity for its weight - though it incurs a significant penalty in printing time. Honeycomb works well with flexible filaments.

TRIANGULAR:

Very similar to rectilinear but with intersections at 60 degrees instead of 90, triangular pattern infill is faster to print than honeycomb while providing many of the same benefits. Triangle infill works well with flexible filaments.

Some slicing software allows the specification of a solid layer intermittently in the print, with a full solid layer every "x" layers. This can be very useful when loads need to be distributed through the X-Y plane, and structures printed this way can be very strong for their weight and material requirements.

If the model is intended to be printed on a multi-extruder machine, it may be possible to specify a different material for printing the infill. This might be done to use a material that can be printed more quickly, to use an extruder with a wider nozzle, or to use a less expensive infill material.

If strength is critical in the design, it can be helpful to select an infill material with an extrusion temperature nearly as high as or higher than the extrusion temperature of the shell material to facilitate bonding. Note that some filaments do not always bond well (nylon and PLA, for example) and that filament composition and properties can vary widely by manufacturer.

PERMEABILITY

For some objects, especially things designed to be used with fluids or gases, permeability will be a consideration. Here, the designer's options are limited, as the majority of the influence will be in the printer calibration and the slicing settings.

The main thing a designer can do to help ensure the print will work as designed is to specify wall thickness and slicing parameters that will work together to provide solid surfaces, reducing permeability.

Where water-tightness is desired, 3-4 layers (shells) in both the "solid layers" (upper and lower surfaces) and the "perimeters" (wall thickness) settings should work, as long as the printer is well calibrated. This is because the small amount of *overextrusion* that is present when the printer is properly calibrated should create very good fusion in the layer 2-3 interface.

To minimize permeability, it is important to design wall thicknesses as exact multiples of the nozzle diameter....For example, with a 0.35-mm nozzle width setting a 0.7-mm wall will print solid, even though the slicer may be set for 1 perimeter and 25% infill. Likewise, a 1.2-mm or 1.6-mm wall should contain no voids if sliced with two perimeters on a 0.4-mm extruder, where a thicker wall may actually exhibit increased permeability due to voids taking up the surplus space.

In the case of stringent permeability specifications, it is necessary to communicate the appropriate slicing specifications and usable nozzle width settings to the printer operator through a readme file or other documentation. This document should include any post-printing treatments such as solvent, paint, or resin sealing if required.

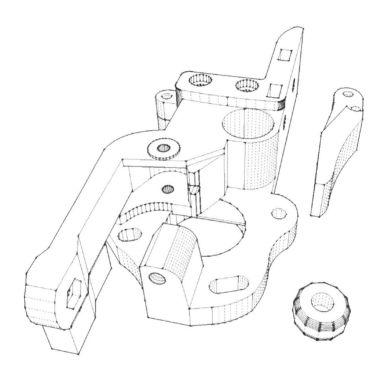

CHAPTER II: DESIGNING FOR THE FFM PROCESS

Designing parts that can be printed reliably using FFM/FFF 3d printing technology.

Apart from the overall form, use, and structure, the main design consideration for 3D printing is printability. Even the most carefully executed design is useless if attempts to print it result in a tangle of extruded filament instead of the piece it was meant to represent.

Sometimes addressing printability consists only of making sure the orientation of the model is correct to minimize the requirement for bridging, overhangs, and unsupported sections. In other cases, a model may have to be subdivided or "refactored" into smaller pieces, support structures may be needed, and bed adhesion may need to be considered.

A well-designed model will be the result of the integration of these factors, and will print reliably on most printers with sufficient strength and the correct form to perform its purpose.

PRINTER LIMITATIONS

Despite its amazing flexibility, 3D printing is limited in the types of structures that can be formed, though less so than most automated manufacturing processes.

The most common limitation of FFM/FFF printing is that of unsupported structures. These are structures that must be formed above the bed, with no structure underneath on which to print them. In these cases, the print nozzle is extruding plastic into "thin air" and support must be provided or the extrusion will droop and the print will partially or completely fail.

When a model needs support for an overhanging or floating part, there are a few options to achieve this. Aside from redesigning or reorienting the model so that support is not necessary, the designer may incorporate structures designed to support the unsupported area. In other situations, it may be decided to let the print processing software automatically design any needed support material in the pre-print processing workflow.

In either case, support material should be just strong enough to print reliably and provide the required platform. It should be be structurally well connected to the print bed or lightly connected to printed layers that it rests on top of. Support material should end just below (one layer) the part it will support so as not to bond to it excessively well, or if actually connected should only touch at a few points or along a line to facilitate easy separation. If support will start on top of a printed area, it should likewise skip a layer first or connect with a limited but distributed surface to prevent damage to the print when it is removed.

UNSUPPORTED STRUCTURES

There are three basic categories into which unsupported structures may fall; completely unsupported sections, overhanging sections that are connected on one end to another structure, and free spanning sections that connect or "bridge" two or more structures.

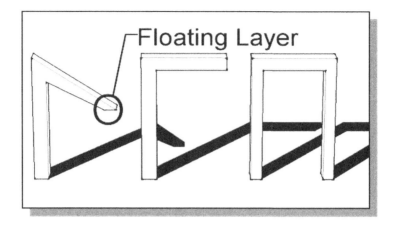

The three types of unsupported structures. On the left, the floating layer will drop to the print bed if it is not provided with support. The middle figure has a significant horizontal overhang. Without support it will (in most cases) droop toward the print bed and will either cause a significant defect or complete failure of the print. The "bridge" figure on the right will usually print without support if the distance between the ends is not excessive. You may notice that all three figures would print reliably and without support if laid down flat on the print bed. (Illustration by author)

When considering unsupported sections in a design, it is often useful to consider changing the orientation of the model or *refactoring* (dividing into separate parts) the design to improve printability. Parts should be oriented on the print bed to reduce or eliminate the need for bridging or support whenever possible.

As an example of refactoring and reorienting a design to reduce support requirements and improve printability, consider a doll house table.

If printed upright, our imaginary table will require support for the tabletop and be very prone to failure when printing. The thin legs and small contact area on the print bed, with the very demanding bridging requirement to form the tabletop, would make it unlikely to print well on most printers.

If reoriented to be printed upside-down, no support would be required, but the legs would still be fragile due to being very high and narrow. Printing them would be slow and breakage would be likely.

If the table were printed upside-down and refactored into a separate table top with sockets for the legs, and the legs printed laying flat on the print bed, the design would print reliably and with no wasted filament used for disposable support structures.

As discussed in chapter one, the horizontally oriented legs would also be printed in their ideal plane for strength, with the layers and perimeters oriented to take advantage of their strongest axes.

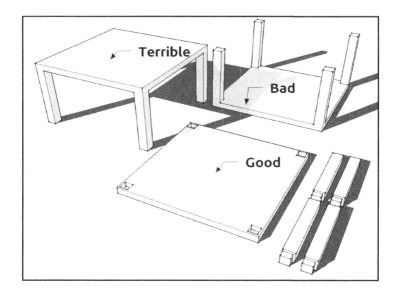

3 different versions of the same model, demonstrating design reorientation and refactoring to optimize printability (Illustration by author - The Zombie Apocalypse Guide to 3D Printing)

Printing speed is also affected by design refactoring and orientation. Our imaginary upright table would print slowly due to the large amount of support material that would also be printed to support the tabletop. The Upside-down table would print somewhat faster without the support material, but will still be slow due to the short extrusions needed to create the tall slender legs layer by layer. The refactored table would print much faster than the other two examples, with all parts in their ideal orientation for printing speed and strength.

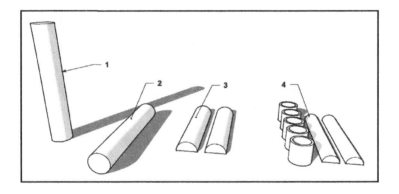

Print orientation and refactoring examples 1-4 for the case study in the following pages. Each of these models will form a dimensionally identical object, but with different strength, printing times, and post-processing requirements. (Illustration by author - The Zombie Apocalypse Guide to 3D Printing)

PRINT ORIENTATION AND DESIGN REFACTORING

In the following exercise, consider 4 approaches to a long solid cylinder, from left to right in the preceding illustration on page 46.

In these examples, our nozzle width is 0.4mm, with a 0.2-mm layer height, 35% infill, and a 60mm/sec print head speed. All of these models are dimensionally equivalent, but each will print with different characteristics.

MODEL 1:

If cosmetic finish is important and strength requirements are minimal, the simple upright cylinder might be a good option if it is wide enough to be stable on the print bed. There could be reliability problems with the print because of the tall narrow profile, but with good bed adhesion or if printed using adhesion aids specified in the slicer, it should generally work. (see section on bed adhesion later in this chapter)

Printing time for model 1 time would be 38 minutes, with 2.53g of filament used.

MODEL 2:

If cosmetic finish is not important, the horizontally oriented cylinder could work and should give reasonable strength and low printing times. Its principal disadvantage is that it may have to be printed using support, and the overhanging underside of the rising surfaces is likely to have a poor finish and will require some post-print processing. Additionally, in some plastics bed adhesion and part warping could be a problem. Finished part strength should be fairly good, and although roundness might be an issue, it would work in many situations.

Printing time for model 2 would be 12 minutes, with 3.13g of filament used. If support were required, time would jump a half minute or so, with 3.2g of material consumed.

MODEL 3:

A generally superior solution to example 2, printing in two halves improves the roundness and finish while eliminating the need for support – improving print reliability and speed. In many plastics it will not require adhesion aids to reliably stay fixed to the print bed. It should be slightly stronger than example 2 as well, but may require adhesive to glue the two halves together. In many cases, this is a good compromise and offers the fastest, most reliable print.

In this example, the print time is just under 12 minutes, with 2.99g of plastic used.

MODEL 4:

This is the strongest version of the cylinder, with a horizontal core girdled by circumferential rings. This will probably be stronger than any of the others if glued, and will have a nice round external profile as well.

At 20 minutes and 3.58g of plastic, it isn't the fastest or the lightest - but it's still about half the time of the other circumferential shell option. Post-printing assembly is somewhat complex, but if properly dimensioned the assembly can be pressed together without glue. In this case, a 0.2-mm gap was left between the core diameter and the rings, which should give an interference fit in most cases, but for optimal strength glue (or solvent welding), would be used as well.

In these examples, in addition to variations in strength and print reliability, printing times ranged from just under 12 up to 38 minutes, with material usage ranging from 2.53g to 3.58g. It is worth noting that a 3x reduction in printing time for the "same" object implies a related decrease in print complexity - and complexity translates well to the probability of print failure.

As you can see, model design can have a drastic impact on printing outcomes even when printing the "same" object. Most of these variables are driven by design choices made when factoring and orienting the model.

BED ADHESION

Bed adhesion is important to consider when designing objects for optimum printability. If a model detaches from the print bed during printing it will almost certainly fail, often spectacularly. Some plastics that shrink dramatically during cooling such as ABS and Nylon are especially vulnerable to adhesion issues, as parts may try to curl away from the bed during printing because of uneven thermal shrinkage.

Small or tall (high Z-axis parts with a small bed contact surface) are especially vulnerable to bed separation from print-process mechanical impact, while large surface area parts are vulnerable to shrinkage, warping, and edge separation.

In the design process there are many things that can be done to enhance bed adhesion and to ensure that the design will print reliably.

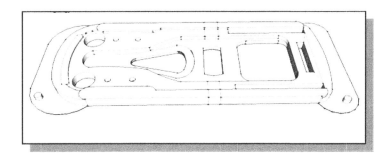

Sub-frame for an engine mount. Note the brim at the ends to ensure good bed adhesion, and the tabs on the brim to help peel it up for easy removal from the bed. (Illustration by author)

Small parts should be designed with a single or double layer 'brim' around the base of the part, which can easily be trimmed off after printing.

Automatically generated skirt or "brim" around a corner bracket to prevent warping and bed separation. Algorithmically generated adhesion structures like this can be specified in the slicing software used by the printer operator. (Photo by author)

This should be around 5 to 10-millimeters wide around the perimeter to provide reliable service, but doesn't normally need to exceed three times the part's width. This kind of simple skirt can also be automatically generated by the printer operator in the slicing workflow.

For long parts or large parts that are prone to curling or warping at the ends or corners, a localized 'brim' feature can be added to keep the vulnerable part of the print properly adhered.

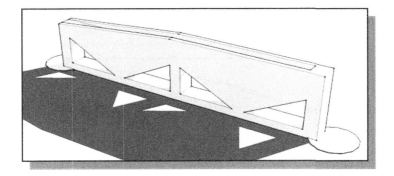

An example of relieving a part to reduce warping stresses. The holes interrupt the tension path that normally causes warpage, while the contact pads at the ends make bed separation much less likely. Note that in-designed anchoring structures such as those shown may be intended for use in addition to algorithmic brim features added in the slicing process. (Illustration by author)

Warping of larger parts is caused by cooling and shrinkage of extruded material as it is deposited. This is not normally a problem for PLA, but with plastics such as ABS and nylon, it can be quite severe. The larger a piece is, the more it will change size as it cools, increasing its tendency to separate from the bed.

Features incorporated into the design to minimize the X-Y tension each successive layer will apply to the layers below it will reduce warping and make your model print more reliably.

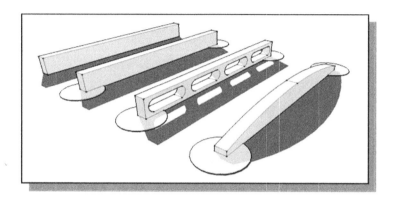

Several approaches to improving bed adhesion through design. From the left, a plain beam, likely to have adhesion issues in some plastics. The second figure shows adhesion pads included to hold the ends down. The next figure is skeletonized to remove warping stress, while the rightmost is tapered downward toward the ends, greatly reducing the lifting moment of any internal tension. (Illustration by author - The Zombie Apocalypse Guide to 3D Printing)

One way of reducing the tendency for parts to warp and separate from the bed is to refactor large parts into multiple smaller, lower profile interlocking objects that can more easily be adhered to the bed.

Stress relieving structures such as holes or slots in long continuous parts will help prevent stresses from building up at the edges of the print.

Careful model refactoring, optimized orientation, stress relief voids and anchoring using "brims" can keep even the most troublesome models firmly attached during printing.

When the print is finished, it will be attached to the bed. The amount of force required to remove the print will be determined by the same bed adhesion factors that were optimized to ensure successful printing. It is important to design objects so that they will adhere to the bed, and yet not be so difficult to remove that damage occurs to the part or the printer during removal.

Most slicing software includes functions for automatic generation of adhesion enhancing features, including brims and rafts. (see photo on page 52) These can be very useful, but since they are often beyond the control of the designer, it can be beneficial to design anchoring features into the model to help ensure successful printing.

OVERHANGING STRUCTURES

Structures designed with overhangs of 45 - 60 degrees from vertical will print without difficulty in most printers (varying primarily due to cooling). This is because a significant portion of each ascending layer still rests on the layer printed below it.

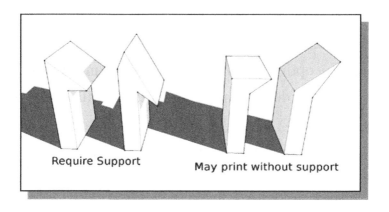

Overhang profiles on the left will not print well or may fail without support. Profiles on the right should print without support on most printers. (Illustration by author)

Lower layer heights improve results, since the horizontal offset of each overhanging layer is smaller with thin layers. Since the nozzle width stays the same, this means that each layer is better supported by the one below it. In any case, each layer must significantly overlap the previous one or they will sag away from the structure, compromising support for subsequent layers, resulting in poor finish quality or print failure.

BRIDGING

Unsupported bridging between objects is possible but is sometimes problematic. Bridging capability varies widely by printer, material and software, but in general, bridging of gaps up to 1 cm is reliable in most materials.

Where bridging cannot be avoided, it is often useful to provide a removable supporting structure for large bridges to obtain better quality results.

The built-in support algorithms in slicing programs may also be used, but it is important to consider the adhesion and removal of the support material in the design.

Bridging gaps over a few millimeters will often result in some sagging or artifacting of the underside of the bridged part. Some professional (or well set up consumer grade) printers may be able to bridge up to 5 cm with minimal problems, but variations in travel speed, extrusion quantity and cooling will all affect bridging performance and finished print quality. Because of this, designers should generally not rely on unsupported bridges of more than a centimeter or so for best-practice design.

Bridging unsupported gaps shares many of the disadvantages of printing over support structures, including poor underside finish of the bridged area, increased probability of print failure and increased reliance on correct printer adjustment and knowledge on the part of the printer operator.

Despite these issues, small bridged gaps can be designed to print reliably with good results, negating the need for support in a carefully designed model. One useful method for reliable bridging that can be used in some cases is to build an arch or converging overhangs instead of an abrupt horizontal bridge. This can be used join two parts while avoiding the vast majority of potential printing issues, as long as there is no need of a clear span below the bridge.

To work at all, bridges must be designed to cross the gap completely on the first layer of the bridge. This means that the lower face of the bridge must be precisely horizontal. Any deviation from horizontal along the first layer of the bridge will cause the bridge to terminate unsupported in midair, and it will probably fail.

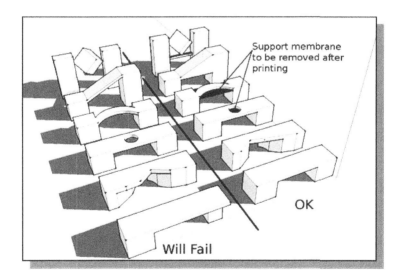

Comparison of bridging structures. Structures on the left will bridge poorly if at all without support. Corresponding models on the right have ideal bridging characteristics. (Illustration by author)

The first couple of layers of a bridge structure must have straight edges and contain as little structural detail as possible. The corners of the bridge platform should be joined point to point, with no angles or additional corners in the bridge itself.

Details such as holes or changes in direction should start on the third or fourth layer of the bridged structure, using a straight, flat bridge as a foundation on which to print these features. Cleaning any resulting "flash" from the print may require some post-print processing, but the result of trying to print detailed structure in the first couple of bridged layers is usually a very messy underside, if not a completely failed print.

A bridge can be used to form the basis for a complex structure "hanging" above it, providing that the bridge starts horizontally and the structure can be scaled out from that initial line of support. An example of this could be a cube suspended from a structure above. If a (removable) bridge is used to support the lower extreme of the cube, it can be built up from there without additional support, eventually joining to the supporting structure above.

Where possible, it is often useful to change the orientation of the print or to design it in several pieces to eliminate the need for bridging or overhangs. This will reduce reliance on the skill of the operator, the features of the printer, and on wasteful and sometimes problematic support material. This will make your designs print faster, cheaper and more reliably.

SUPPORT STRUCTURES

With the exception of bridges and allowable overhangs, every printed element in each layer must be supported by a layer beneath it - either the bed, lower layers of the print, or a temporary "scaffold" of support material.

A print layer should not be started or ended in "thin air". Overhanging structures pointing in a horizontal, nearly horizontal, or downward direction cannot normally be printed without some kind of support.

Three otherwise identical parts printed in different orientations. At left, a 'T' printed upright with support, showing surface imperfections where the support was removed. On the right, the 'T' was printed upright without support causing the unsupported sections to fail. The center example was printed flat on the bed. (Photo by author)

It follows then, that a T-shaped figure cannot be printed erect without support. It should instead be printed lying flat or perhaps upside-down in order to print without defects. Alternatively, support structures could be specified to support the ends of the T from the build platform. This would work, but will consume additional plastic, take additional time, and the bridge between the primary pillar and the support pillars will be likely to have minor imperfections.

In cases where it is not practical to reorient or refactor the model to eliminate unprintable overhangs or bridges, disposable support structures may be incorporated into the design. When it is not desirable or deemed necessary to design support structures into the model, the slicing or printing software can be allowed to generate support structures automatically.

The characteristics of automatically generated support can vary depending on the particular software and settings used by the printer operator.

Some software stacks do an excellent job of generating support structures, but if your model is to be printed by someone using an unknown printer with unknown software, it is often best to minimize or eliminate the need for automatically generated support structures through careful design.

Generally, algorithmically generated support structures fall into two categories – prismatic "accordion" support and branching, "tree" support.

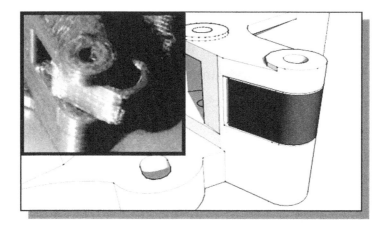

An example of designed-in support (dark gray). Note the small separation between the support and the supported tab. This support structure will print and support more consistently than most automatically generated support. With a gentle press, the support snaps out cleanly and easily in one piece. Designing in support instead of using automatically generated structures is usually not necessary for one-off prints, but can be very practical for things that will be printed many times like this extruder assembly. (Illustration by author - The Zombie Apocalypse Guide to 3D Printing)

Prismatic support "grows" directly up from the print bed from a simple "shadow" footprint to support all overhanging or floating surfaces. It is typically a sparse square, triangular, or zigzag design, with elements spaced just closely enough to facilitate easy bridging while minimizing the material used and the structural strength of the lattice.

Branching support "grows" up from the base in narrow towers, branching out into limbs to offer only the minimum support required to the structure.

Branching support, though very common and effective in resin printers, is less reliable in the FFM process due to the relative fragility of the tree-like construction. This is somewhat less of a factor for printers where the bed moves only in the z-axis (due to not shaking the print), but even then an unfortunately misplaced blob of plastic can easily cause print failure due to breakage or bed separation of the support tower if the hardened blob is struck by the print head.

Because of the relative material efficiency and potential speed gains, branching type algorithms are likely to see improvements that will hopefully bring their reliability closer to that of prismatic support.

MANAGING AUTOMATICALLY GENERATED SUPPORT

When automatically generated support must be used, careful management of support requirements can improve the printability of the design. Unfortunately, this can be very difficult from a designer's perspective without having specific knowledge of the software stack and support geometry with which the model will be prepared. Though sometimes a difficult challenge, it is still possible to minimize any negative impact that support structures may have on the printability of the model.

Without direct control or knowledge of the pre-print process, the first step is of course to mitigate the need for support structures. When the limits of practicality for this approach have been reached, consideration must be given to the likely distribution of any generated support material.

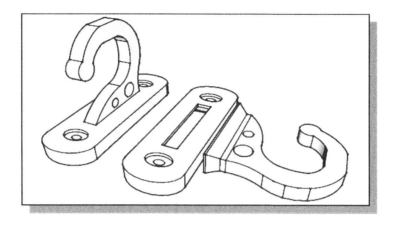

Otherwise identical hooks, one for one piece printing, the other to be printed in two pieces. The one piece model will require support material to print. The two piece model will print faster, use less material, print more reliably, and be much stronger than the one piece model. Although both will require some degree of post-print finishing, the two piece model can be merely assembled without troublesome removal of support material. (Illustration by author)

It is critical to ensure that no support material or model parts will become trapped during printing. Poorly managed support requirements can cause automatic algorithms to generate support material that will be difficult or impossible to remove without damaging the model.

This problem can be anticipated by looking for weak or difficult to access structures in the design that may become trapped in support material, as well as identifying areas where print structures will cause any required support material to be difficult to remove.

One way to mitigate this trapping problem is to build break-away dividing structures into your model so that the support material will be subdivided in a predictable fashion, easing removal. This can take the form of a thin (typically 2-4 nozzle widths) wall, anchored to the pint bed with a suitable base, that extends up to within a layer or two of the structure that will require support. Automatically generated support will then be divided by this disposable barrier, easing removal. If support barriers must rise up from print structures, they can be given an inverted 'V' or saw-tooth bottom, so that they can be easily removed from the finished piece.

Especially where support structures will be tall and narrow, it can be useful to provide a thin base of one or two layers to which the otherwise fragile support material can adhere for stability. This is important because print failure is likely to occur if the support material separates from the print bed, and some support algorithms may not automatically provide this base.

If the structure to be supported is small and will begin far above the level of the bed, it can be useful to build a platform (tower) under the anticipated support material, as some algorithms will not adequately reinforce very tall and narrow supports. Inadequately strong tall support structures may break off during printing, causing print failure.

If the printer being used is equipped with multiple extruders, it is possible to use different materials for structure and for support. In these cases, a dissolvable support structure can be specified. This will be then washed away with a solvent after printing, providing optimal support with minimally invasive post-print finishing. Dissolvable support structures can be design-specified, automatically generated, or in some problematic cases made by subtracting the model from a solid to create a 100% supporting structure. *(Also see multi-material printing, page 146)*

Dissolvable support can make otherwise unprintable models relatively simple to design, permitting completely entrapped elements that require support to print. Dissolvable filament choices include High Impact Polystyrene (HIPS, dissolvable in limonene) and some water soluble PVA variants. Dissolvable support enables the printing of very complex objects in any orientation, at the expense of printing time, material cost, and the requirement for special post-print processing.

There are improvements in support-generation algorithms being made all the time. Skin-frame support[1] and infill, for example, shows great promise in this regard and may soon find its way into commercial and open source model processing software.

Failed support structure. Here, there was no adhesion pad in the design to print the support material on, and some of the material has separated from the bed. A base provided by the designer for anticipated support material would have prevented this print failure. (Photo by author)

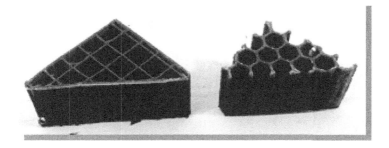

Two types of automatically generated support structures. (Photo by author)

Two different support algorithms used with the same model. Note that there is an adhesion pad provided by the designer for the support material to print up from. (Photo by author)

Model with preliminary support removal completed. This model could have been printed as multiple stacked parts and no support would have been necessary. The print would have required some post-printing assembly, but would not require support removal – saving time and material, while providing a better-finished print. (Photo by author)

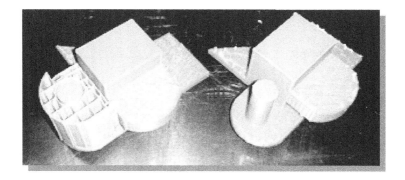

Trapped support material. On the left, support material has become fused on the round column. Breakaway support barriers or the use of a different support algorithm (model to the right) could mitigate this problem, making the removal of support material much easier and less likely to damage the print. Many slicers also allow for adjustment of the clearance between print structures and support material. Specifying a larger clearance can ease removal at a small cost to support effectiveness. (Photo by author)

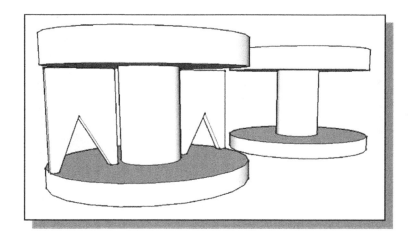

Breakaway tabs to divide or limit support material. Here, two tabs divide the support material that will need to be generated for the part on the left. The tabs are intentionally designed thin and do not quite reach the base of the upper disk, to facilitate easy removal. In this example, the barriers are anchored with an inverted 'V' because they are anchored to an upper surface of the model rather than the bed. (Illustration by author)

This illustration shows a toolpath cross-section (slice) 8-mm up from the base of the model shown in the previous illustration, detailing the support structure (light gray) and the model cross-section (black). Notice how the support structure on the left is divided by the black tabs. This support material will be easy to remove in two halves without risking part breakage, even if a very durable support algorithm is selected. The support material for the part on the right will completely surround the axle, trapping it. In practice, it might be better to refactor this assembly as two disks and a separate axle, requiring no support to print. (Illustration by author)

*1: ACM Transactions on Graphics
(Proc. SIGGRAPH Asia), 32(5), Article 177: 1-10, 201*

CHAPTER III: DESIGN FOR STRENGTH

Optimizing design for desired structural characteristics and reductions in material usage.

Parameters such as flexibility, weight, rigidity, and durability are determined as much by design as by material selection. By working within the limitations of the manufacturing process, 3D printing allows the designer to fine tune the structural characteristics of the material to a significant extent, adding special features to the object where needed and creating voids or empty sections where desired.

3D printed structures can be optimized in ways that might be very difficult for other fabrication techniques. Due to the extensive automation involved, this type of complexity can be achieved at near zero marginal cost. This unique feature of additive manufacturing frees the designer to add functionality, customization, and refinement that might be uneconomical to contemplate in other manufacturing processes.

A BRIEF PRIMER IN BASIC MECHANICAL DESIGN

Author's note: Functional objects will often be subjected to significant forces that they must endure without damage. Mechanical design deals with the principles involved in mitigating these forces to ensure that parts perform reliably and as expected. This field is a complete engineering discipline unto itself. Here I provide only a 10,000-meter flyover view of some of the very most basic principles, those which every designer will find critical for functional designs. The treatment here is dramatically simplified and much is omitted entirely.

FORCE VS. STRESS

It is easy to fail to differentiate between force and stress. Fortunately, it is equally easy to understand the difference, and this is a difference that it is imperative to grasp as a designer of functional objects.

Static force describes a fixed external influence on an object, and is usually expressed in units of mass, such as kilograms, pounds, ounces, or grams. Dynamic force is sometimes described in terms of acceleration with the mass implied - as in meters per second per second, or as expressions of force over time, as in newtons.

In our examples, we will only be examining static forces, those forces that can be expressed as static units of mass, even though many static forces are actually expressions of dynamic forces such as gravity or acceleration.

<u>Stress</u> is the expression of a force exerted on a material object, and is usually expressed in terms of force per area. For example, if we have a 1kg brick balanced squarely on a 1cm cube, we could say that the cube is experiencing a stress of 1kg per square centimeter. As long as the compressive strength of the material of the cube is more than 1kg per square centimeter, the cube should not deform significantly.

The important distinction to make is that force is an external influence, while stress is the structural effect that influence has on an object.

TYPES OF STRESS

Mechanical Design deals with objects and the stresses produced by forces applied to them. Of the different types of applied mechanical stress, we will look briefly at compression, tension, shear, and bending stresses.

It is worth noting that rarely will one encounter "pure" stresses in actual practice. In most cases, the stresses involved will be a combination, but it is usually possible to identify the most significant stresses and design for those – most of the time, the minor stresses will be accommodated.

Apart from the stresses themselves, I will give an incredibly brief overview of the most critical stress factor for the designer – stress localization.

TENSION:

When a part is stressed in tension, it is being stretched by two opposing forces that attempt to pull it apart. The overall force is the same in all planes perpendicular to the direction of force, and the total stress at that plane will be a function of the amount of force divided by the cross-sectional area of the part.

The cardinal rule of design for FFM manufacturing is to avoid tension in the Z-axis, as tensile strength in this direction is much less than X or Y axis strength when printing in most materials.

COMPRESSION:

When a part is in compression, it is being pressed upon by two opposing forces that attempt to crush it. As with tension, the overall force is the same at all planes perpendicular to the direction of force, and the total stress at that plane will be a function of the amount of force divided by the cross-sectional area of the part.

The main design feature that can improve compressive strength of 3d printed structures is to design the part so that there are ample skin structures oriented to directly bear compression loads deep into the part. This can be accomplished with the overall part shape, or by adding strengthening holes to the load bearing surface.

SHEAR:

Shear stresses are created by two opposing yet misaligned forces. Imagine a stack of typing paper two inches high. Now, imagine pushing the top of the stack to the left while holding the base of the stack to the right. The stack will deform by "shearing" to the left from the force applied to the top of the stack. Shear stresses are distributed along planes in the direction of the applied forces, distributed generally between them, rather than the perpendicular orientation seen with tension and compression. Shear stresses are always accompanied by localized compression and tension stresses within the structure.

The situation to avoid with shear stress in FFM manufacturing is having the shear direction on the XY plane, as shear strength in between XY planes tends to be lower than on the XZ or YZ planes.

BENDING:

When a part is subjected to bending stress, it is acted upon by at least three forces. Typically, two primary forces act against a third force, like a person standing on a plank spanning two blocks, or on the end of a diving board. The effect of a bending force is to deform the part, often in complex ways depending on its structure. A bending force subjects different parts of the structure to varying tension, compression, and shear stresses.

Bending stresses are very complex, so there are few hard and fast rules for 3d printing – but in general, a part will be stronger if the flexion from bending loads happens mostly along the X or Y axis, avoiding heavy interlayer shear stresses that might cause layer delamination, while especially avoiding any interlayer tension stresses if possible.

Torsion:

For the purpose of (over)simplification, one can think about torsion stresses as a rotary version of shear stress. An example for torsion stress would be the stress on a shaft, with a turning motor on one end and a wheel on the other. If the wheel is held, and turning force of the motor exceeds the torsion strength of the shaft, it will break.

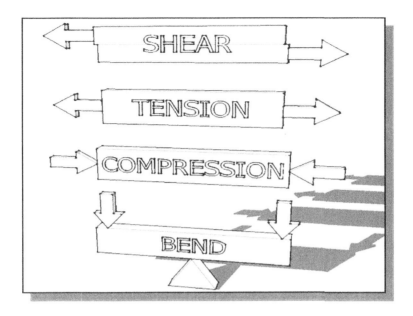

Some of the many types of stress that a part may be subjected to. (Illustration by author)

Torsion failures occur either along the same plane as the torsional force, or as tensile failure perpendicular to that plane. Because of this, we can reason about torsional loads in 3d printing as if they were essentially shear loads in the plane of the torsion force, with tension loads perpendicular to it. Either way, the rule of thumb would be to orient the torsion force off of the XY plane for shaft-like objects.

One might intuit that torsion and bending loads are basically composite constructs of compression, tension, and shear loads. This view, while very simplified, is not entirely without merit. In this model of thought, there is only compression, shear, and tension, while other mechanical forces such as bending and torsion are composites used to simplify aggregating complex combinations of basic forces to more easily reason about them. This model can be very useful for reasoning about how a complex load might be broken down to localized tension, shear, and compression stresses in the design. When visualizing your design in this paradigm, try to imagine where it will be in compression, which sections will be pulled apart, and where the part may try to shear.

STRESS LOCALIZATION

It is imperative to note that in introducing the fundamental tension and compression stresses, I proposed that the overall force is the same at all planes perpendicular to the direction of force. I also said that *overall* material *stress* at that plane will be a function of the amount of force divided by the cross-

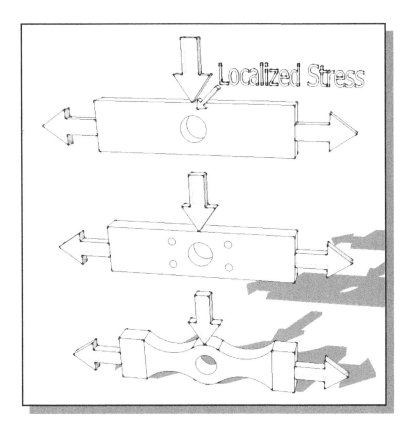

Here, a beam is weakened by a penetration. Stresses are localized at the thin edges of the hole, and the beam will be weaker than necessary. The bottom example trims away a lot of material and causes the loads to be more evenly distributed over the length of the beam, as well as bringing the stronger skin structure closer to an ideal alignment to resist the cumulative forces. The middle example uses additional holes to spread the area of reduced cross-section over a larger area, as well as implementing additional penetrations to the skin surface to carry internal loads to the outer skin. (Illustration by author)

sectional area of the part. While strictly true, this can be very misleading, and the clarification of this point leads to the discussion of one of the most important concepts in structural design – stress localization.

Dramatic changes in the shape or size of this cross-section from one perpendicular "slice" to the next will tend to "focus" a disproportionate amount of the overall stress on a relatively smaller *part* of the cross-section, rather than it being evenly distributed. This can cause localized failure of the part under stress, leading to damage or catastrophic failure.

These heavily stressed areas are described as areas of localized or concentrated stress, as the overall stress on the part tends to become focused on these points. Design decisions play a critical role in strength and durability by identifying and mitigating these stress localizations.

The basic principle of reducing stress localization is to smooth transitions in shape so as to distribute the stresses over a larger area.

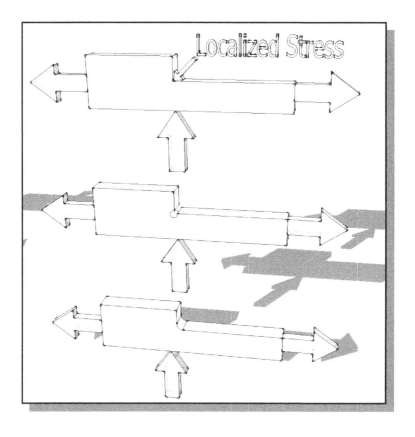

Here, a beam is weakened by an abrupt change in cross-section. Stresses are localized at the corner, and the beam will tend to crack at this stress concentration. The bottom example provides a small rounded fillet, spreading out the concentrated stress over a larger area and transferring it to the adjacent surfaces. The middle example is useful when a fillet cannot be used and merely spreads the stress out over the circumference of a round cut. This second method is especially useful when two sides of a part must converge at an acute angle. (Illustration by author)

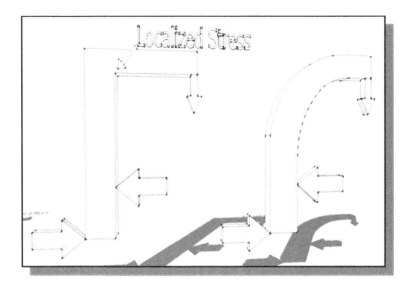

Here, two crane structures may be compared. In the example on the right, a sharp angle is replaced by an arc, saving material. The curved crane will be stronger because the structure is more closely aligned with the direction of forces, and there is no sharp angle to cause dramatic stress concentrations. (Illustration by author)

Because strength is concentrated in the shell with FFM printing, this is an especially effective strategy for functional print design.

There are some other strategies that are less intuitive but achieve the same goals, including trimming other areas of a part so that stresses are shared over a larger region, aligning the axis of a part along the primary stress vector, and utilizing the skin strength of holes to redistribute loads over a larger area or to adjacent perpendicular faces.

CORNERS, INTERSECTIONS, BOSSES, AND RIBS

Although one of the many advantages of 3D printing is freedom from the constraints of molded parts, basic parametric rules used in molding plastics are still part of 'best practice' design.

With 3D printing, it is possible to create large monolithic parts without being limited by the endemic shrinkage and hollowing problems of cooling injection molded parts. Though these molding problems can be safely ignored, FFM printing comes with its own set of cooling challenges such as warping and bed separation. These problems have design remedies distinct from common practice in injection molding. These issues typically do not categorically constrain design freedom and are discussed separately in the section "Bed adhesion" within Chapter II.

After dealing with the process specific idiosyncrasies involved with printing in the FFM process, the mechanical design considerations remain.

Although monolithic blocks with lightweight or solid internal structures are practical with FFM/FFF, it is often desirable to construct minimized structures to accomplish functional goals, preserve flexibility, conserve weight, or reduce material costs and shorten printing times. When designing these types of models, features such as wall thickness and corner radii become important considerations, as do the details of intersections, bosses, and ribs.

In general, the best practices of injection molded parts apply here, with some modifications and caveats on feature thickness.

To ensure void-free construction, the width of thin vertical (Z-axis) design features should not be less than 2 nozzle widths wherever possible, and should not exceed 2x the number of specified perimeters (for example 4 nozzle thicknesses, if 2 perimeters are to be specified in the slicer). If they need to be stronger, they should be made *significantly* wider, so that they will be wide enough to accommodate proper infilling. (also see *Chapter VIII*).

Use uniform wall thicknesses where possible, with a generous radius at corners to minimize stress localization. The inside corner radius should be at least equal to the wall thickness when possible. Use ribs instead of solid blocks to reinforce or gusset wall structures, using rib thicknesses equal to the wall structures they reinforce. Use radiused intersections wherever possible.

Unlike injection molding, thickness transitions present no fabrication challenges with 3D printing.

Where structural integrity is important, wall thickness transitions should be tapered and be at least three times as long as the transition difference, more if possible. For example, a one-millimeter wall thickness change should ideally take place over 3-millimeters or more, unless the part has a continuous taper. This will reduce stress localization and create a stronger, lighter part that prints faster and more reliably. Using curved instead of angular transitions can further increase strength while improving printing reliability by reducing acceleration loading on the extruder during printing.

Keep in mind that infill algorithms may leave certain wall thicknesses completely hollow, depending on the infill ratio. (also see *Chapter VIII*).

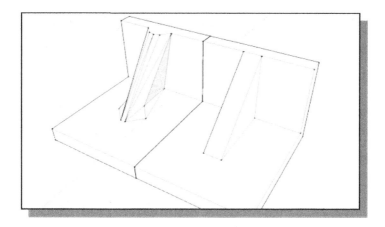

One of the advantages of additive manufacturing versus other manufacturing methods is the ability to more freely optimize structural features. The gusset on the left, while using about the same amount of material, concentrates its strength along the critical edge and spreads out its Z-loading over a larger plane. It will be much stronger (and more resistant to delamination) than the plain gusset, but its pinched shape would cause complications for injection molding. (Illustration by author)

Ribs and bosses are often lightened or improved by gradual tapering, so the same amount of material can be used to spread the load over a larger footprint.

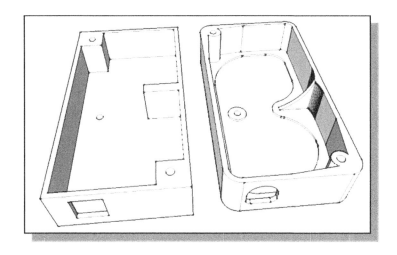

"Bad" (left) vs. "good" (right) design choices for two functionally identical boxes. The box on the right will print faster and more reliably while being more durable, flexible, and lightweight. The box on the right will also have a smoother, more finished appearance due to a reduction in printing process artifacts caused by sudden acceleration changes of the print head. (Illustration by author)

In all cases, it is best to keep in mind the anisotropic nature of the FFM process, minimizing tensile or cantilever loading of high Z aspect ratio structures wherever possible.

LAYER FUSION

Layer fusion, while being primarily a production rather than a design issue, must nonetheless be considered by the designer. Although most critical in the Z-axis, it is best to minimize the dependency on fusion wherever possible, as improper fusion is a very common printing problem. Because of this, designers should maintain an awareness of the degree to which they are depending on the complete fusion of filament, both in the X-Y plane and in the Z-axis.

To improve strength characteristics, models should be printed in an orientation that confines significant loads to the X-Y plane of the printed piece, and skin structures oriented with the anticipated direction of force.

In more complex assemblies, good design can help reduce layer fusion issues by factoring a design into components that depend less on Z-axis tensile or shear strength. This goal can be accomplished by using multiple pieces, each with optimized orientations. Components may interlock, be glued, or connected with fasteners in the final assembly.

If fasteners such as screws or bolts will be used in the design, they can be leveraged to compensate for inter-layer weakness in the structure by using them to hold the Z-axis in compression.

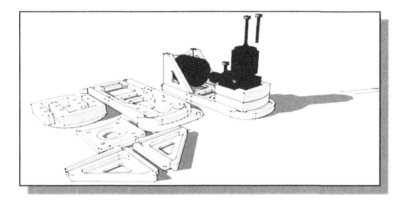

A drone Auxiliary Power Unit (APU), showing how the engine mounting bolts hold the stacked components in tension. giving them the required strength to hold the engine and generator (black) in place despite the severe operational environment. Note the bed adhesion enhancing features of the parts as shown flat for printing. (Illustration by author)

Reducing dependence on fusion on the X-Y plane consists of orienting the skin structure to bear expected loads along the surfaces of the object. One way to think of this is to imagine that the printer is going to print the object out of sticky string. The string sticks together, but not nearly as strongly as the string itself is. Any loads would obviously have to be oriented along the lay of the strings, or the strings would just be torn away from each other.

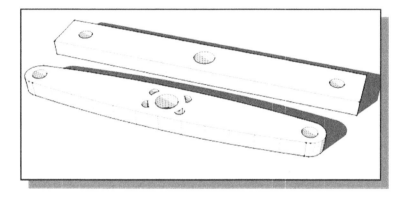

Two functionally identical beams. The lower one transfers the loads to the structural skin more effectively, avoiding stress localizations and shear loading. It also uses slightly less plastic and prints faster. (Illustration by author)

In practice, this means carefully considering the directions of all major forces, orienting or creating additional surfaces in the direction of strain, and transferring these strains effectively using rounded sections at the loading points. Rounded sections avoid translating tensile loads into shear loads at the attachment points, and reduce stress localization.

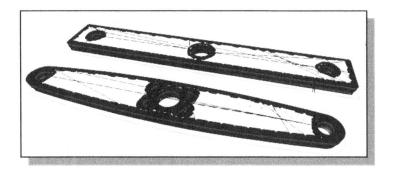

Internal structure of the beams in the previous illustration. Notice how the fastener holes are bonded directly to the skin in this design and how the small holes around the center axle are used to create a reinforcing structure. (Illustration by author)

CROSS-SECTIONAL CHARACTERISTICS

When designing parts with significant vertical height, it can be useful to imagine or examine the Z-axis cross-section at various points. This is especially important if the object in question will be subject to significant loads.

For example, transitions from horizontal to vertical surfaces should be made with rounded intersections to minimize stress localization. Simple gusseting is often sufficient to provide any needed reinforcement, and the material or weight penalties for this are minimal.

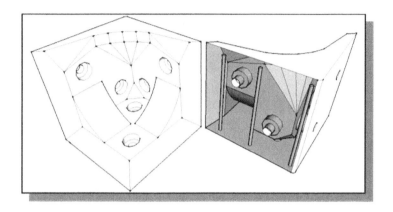

A corner bracket, with the back cutaway on right. X,Y, and Z cross sectional transitions are managed to minimize stress concentrations and resist layerwise delamination. Tiny 2-mm tubes to the bottom help to "spot-weld" the Z-axis, increasing layer bonding strength near the critical mounting holes that will hold the screws when the piece is in use. (Illustration by author)

Designing holes through the Z-axis can provide reinforcement for monolithic structures, creating reinforced columns within the otherwise relatively weak infilled areas and helping to "spot-weld" the layers together. This works best with multiple small holes rather than fewer larger ones, because when small holes are printed, heat is concentrated in the printing area and interlayer fusion is locally improved.

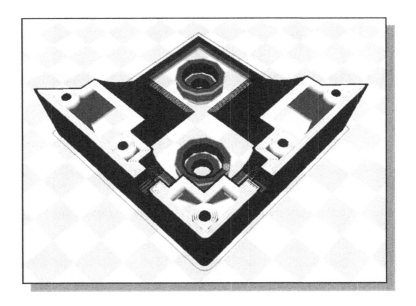

The corner bracket from the previous illustration, showing slice detail. Note how the small holes form solid columns, reinforcing the fastener holes and the Z-Axis. (Illustration by author)

Strengthening holes can be especially useful when significant compressive or tensile loads will be supported in the Z-axis. The resulting tubular columns enhance the ability of the structure to resist shear loads and delamination by providing multiple bonding points between layers.

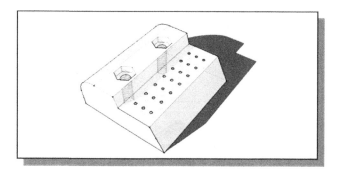

A slide stop bumper with reinforcing holes designed in to give it more hardness and strength. These holes will form solid column structures, similar to the corner bracket shown in previous examples. This is a simple way to reinforce the Z-axis that is almost always available to the designer. (Illustration by author)

Reinforcing holes can also be utilized with partial penetration to enhance rigidity within a limited area, as with a bonding pad on the upper surface of a flat area which might otherwise be easily torn away from the underlying infill.

An example of reinforcements engineered in this way would be a dense area of penetrations extending 2 to 5-millimeters into an otherwise larger part, to provide a reinforced layer to bear and distribute anticipated loads.

The slide stop from the previous illustration shown by CURA in slice view. Here you can see that the reinforcing holes form a nearly solid section, significantly reinforcing the part. (Illustration by author)

This reinforced face would then have the strength characteristics of a thicker surface without the attendant weight, printing time, or material use penalties.

Small reinforcing tubes work best when oriented along the Z- axis, but horizontal ones will also provide significant reinforcement. This is especially true if they are relatively small (<2-millimeters). If large X-Y oriented tubes are needed, a triangular form at the upper part of the tube, converging at 45 degrees or less from the vertical, will ensure that support problems are not encountered during printing. Square horizontal tubes oriented as "diamond" forms work well for this.

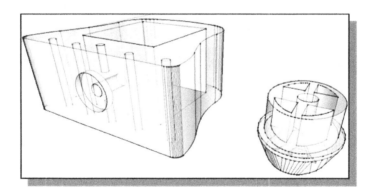

Furniture end plugs for use with tubular steel sections. Note the small reinforcing holes in the Z-axis of the larger foot for the shelving unit (printed on its side) and the reinforced design of the plug for the oval tube. Both pieces will be stronger than if they were designed as prismatic solids - unless a very high infill was used, consuming unnecessary material. (Illustration by author)

DETERMINING FAILURE AND WORKING LOADS

If an item must be rated for a specific load, its failure load should be determined as part of the prototyping process. The failure load is the load at which the item breaks completely, begins to deform so that failure is imminent, or deforms so that it can no longer perform its intended function.

To determine the working load limit, a test to failure of at least 5 samples should be performed. Obviously, this requires access to the material and printer with which the part will be made. The printing environment may have an effect on build strength, so factors such as filament humidity, environmental temperature, and the like should also be controlled in the printing process. An enclosed printer in a temperature controlled environment is a minimum requirement for this type of production.

The test prints should utilize the same printer, printing file, software settings, print speeds, and filament batch for each test model. These must also be used in production for test data to be meaningful. If the test will be formalized into a load rating, the failure loads of each test should be recorded and the mean and standard deviation calculated.

To test the printed object, it should be placed in actual use or in a simulated use environment, then loaded with progressively higher / heavier loads until the part deforms or fails. At least 5 identical tests should be performed on 5 identical models for the data to be statistically meaningful.

The failure test should closely approximate the use environment. If there will be motion, the *worst case* motion should be incorporated into the test as accurately as possible.

The calculated minimum failure strength (MFS) of the printed item can be considered to be approximately the mean failure load minus 3 times the standard deviation of the tested failure loads. This will provide a confidence level of 95% for the MFS (minimum failure strength) rating. If a higher confidence level is required, more test samples must be used.

A way of estimating a nominal working load limit is to divide the calculated minimum failure strength (MFS) by a safety factor of 3 for static, non-moving items, using a factor of 4 to 10 for moving or lifting items. For especially violent motion, safety factors may need to be even higher.

If the item will be subjected to long-term heavy loading, testing should ideally be done out to twice the expected load times to test for progressive failure. Any notable or *progressive* deformation should be interpreted as part failure.

It is, of course, inadvisable to use FFM manufactured objects in life safety critical roles, and this section should not be taken as an endorsement to design or manufacture items that could foreseeably lead to injury or death in the event of their failure.

CHAPTER IV: DESIGN PARADIGMS

Several design solutions for 3d printing.

In design, there are at least as many paradigms as there are designers, but some overarching concepts bear outlining as useful guidance or inspiration.

Perhaps the most obvious design paradigm is the simple solid. Here a single, monolithic piece is designed that meets all the design criteria for the finished object. This is usually best for simple forms with facile mechanical requirements - but very sophisticated parts can also be made this way with the use of internal reinforcing structures and careful attention to printing-related parameters.

The primary weaknesses of this approach are the lack of tensile strength along the "grain" (Z-axis) inherent in FFM printing and the difficulties of forming overhangs or bridges with their attendant costs in time, material, and print reliability.

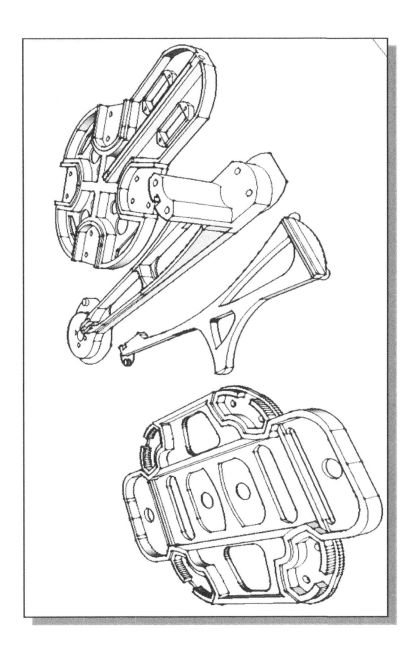

(Previous Page) *Multi-paradigm design, using solids, stacked design, beveled joinery, and snap fit parts to construct a drone chassis. Four of the arm assemblies shown bolt between the top and bottom plates, with the arm reinforcement/landing gear braces snapped into place. (Illustration by author)*

A variation on the solid design paradigm is the stacked design. The stacked design seeks to overcome Z-axis weakness by incorporating one or more fasteners through the Z-axis into several (or sometimes only one) stacked parts. This has the additional advantage of reducing support requirements and enabling perfect bridge structures by fastening the overhanging part to the top of the lower parts in final assembly and giving the designer the option to refactor and reorient parts for maximum strength and printability.

Fasteners used in stacked designs may include bolts, screws, pins, or various printed retainers such as rings, printed pins, or other mechanisms.

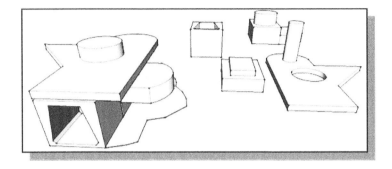

An example of two design paradigms. The complete model on the left will have to be printed with support and includes a brim (wide flat base) to ensure support integrity. The same model can be printed in four parts (shown to the right), each ideally oriented to print reliably and without support. The four parts can be assembled in less time than it will take to remove the support material from the model on the left. The model printed without support will print faster, with less material and more reliably than the one on the left. Note the orientation change of the hollow block to facilitate support-less printing.. (Illustration by author)

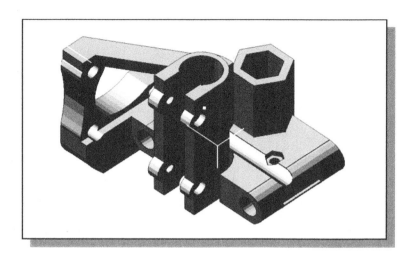

Example of a complex solid part - in this case, an X carriage motor mount for a Prusa Mendel printer. Some parts of this design will be fairly weak in the Z-Axis tensile mode. (Illustration by author, original part design by Josef Prusa)

Where Z-tensile strength is not critical, stacked structures may be glued together. in this case, it is useful to use alignment holes and printed or manufactured pins to ensure correct alignment while gluing.

Elaborating somewhat on the stacked and fastened approach, a completely multi-orientation build often offers the best combination of strength, light weight, printability, and low non-printed fastener count.

A stacked structure, held in compression by bolts. In this case, an engine bed for a drone APU. This application requires high strength, rigidity, and vibration resistance. The compressive strength of the stacked layers is complimented by through bolting at the motor mount. (Photo by author)

In this approach, a complete assembly is printed in multiple components, each one with optimized print orientation, so as to minimize any undesired anisotropic structural effects. Tab and slot, dovetail, or snap fit joinery is used to join the components into a completed assembly, and joints may be glued where needed. This leverages the strengths while minimizing the weaknesses of the FFM process, but carries the disadvantage of increased design overhead and significant post-print processing and assembly before a piece is ready for use.

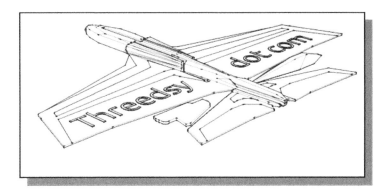

An example of a glider design that utilizes a combination of 'design by slice' and conventional geometry techniques. Here, the edges of each slice are in some places defined by the designer and are clearly visible in the model in some areas on the wings. This design must be sliced at approximately 0.20-mm to perform as intended. (threedsy.com)

A design paradigm with intriguing possibilities is the "origami" method of design. With origami design, flat structures connected by membrane (living) hinges are folded and glued or fastened into their final form.

Models designed within this paradigm have excellent rigidity and strength, with low material expenditures. Although post-printing assembly is required, integrated hinges, formers, and guides can largely automate the assembly process.

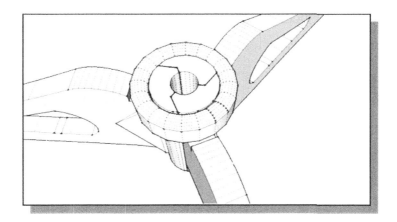

A tripod. The three legs print flat on the bed, and are held tightly together by rings on the top and bottom of the center section. This design snaps securely together without glue. (Illustration by author)

The main limitations to the origami approach are the restrictions it tends to place on design flexibility and the current lack of automated tools to facilitate the design process.

Hopefully, origami algorithms will someday be integrated into common design tools, to enable a solid to be automatically decomposed into an origami representation. Solidworks *(Dassault Systèmes)* has design tools that provide some of this functionality for working with sheet materials.

In the absence of such design tools, manual design is possible and can yield very good results with care.

Hinges should be kept to 2 layers (or more, if it will be heated to bend, or the material is very flexible) and mating vertices and hinge widths used to constrain the bend angles.

Panel thickness should be kept to a minimum and honeycomb, triangular, or cubic infill can be used to good effect here for stiffness. It is often possible to reduce the center of the panel to membrane (2-4 layer) thickness if the edges are properly framed. In this type of construction, the center 'web' of the panel will be loaded only in X-Y tension.

So far, trial and error seems to be the only way to achieve reliable results with origami design, but with the development of more capable design tools this method could yield excellent productivity and functional qualities.

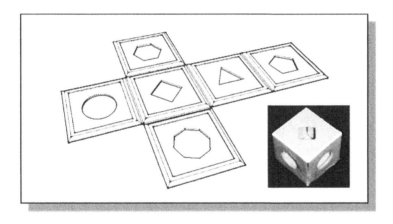

Origami construction. This 20-mm ABS cube prints at under 2.5 grams and supports 139kg (307lbs) prior to failure, even with incomplete gluing of non-hinged vertices. (Illustration by author)

CHAPTER V: DESIGNING FOR FIT

Designing parts that fit together properly after printing.

Note: Some slicing software, including Slic3r at the time of writing, includes internal compensation for overextrusion errors so that dimensional allowances for overextrusion may not be necessary to include in a design. The designer should note if such corrections are incorporated into a model to avoid double compensation for dimensional error.

The figures given here are typical for open source FFM printers with average calibration accuracy and settings, <u>without</u> automatic dimensional correction in the slicer. They represent a starting point or rule of thumb only. If you will have access to information about the specific accuracy or characteristics of the printer and software for which you are designing, that information should be used preferentially.

When printing complex features such as holes, tabs, or interlocking structures, even a properly calibrated FFM printer typically has some dimensional error due to nominal over-extrusion. Of course, the best way to measure this deviation is by sampling, but the designer does not always have access to the target printer, and even if she does, it is useful to have a starting point that will at least be close.

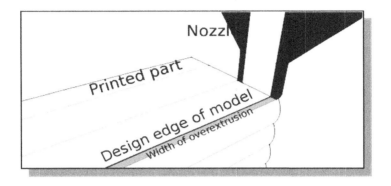

Diagram of printed part construction, showing overextrusion. (Illustration by author)

One of the characteristics of the FFM process is a small amount of overextrusion. This causes printed parts to be nominally larger in the X-Y plane than the design would indicate. As a rule of thumb with most open source FFM printers, the designer should expect around 0.2-mm of overextrusion at each edge, so a part designed at 20-mm wide might be expected to print at 20.4-mm width. Overextrusion dimensions will vary based on the slicing software, nozzle width, and layer height. Some slicers automatically compensate for estimated overextrusion errors.

Overextrusion must be accounted for in the design process (or by the slicer) to ensure that the printed size of an object is correct for its application and to allow the fitting of interlocking parts or fasteners. Using the 0.2-mm rule of thumb for a manual overextrusion allowance, holes should be designed 0.4-mm wider than free passing fasteners, and joints given 0.2-mm additional clearance *from each printed edge*. Using 0.2-mm overextrusion allowance, a printed pin must be 0.8-mm smaller than the design dimensions of a printed hole it will pass through (0.2-mm x 2 sides of the pin and 2 sides of the hole).

In general, overextrusion on a properly calibrated printer (when no compensation is done in the slicing software) tends to be about 0.2 – 0.6x the nozzle width, with taller layer heights tending towards the upper edge of that scale. If the print is to be done on very large nozzles (>.6-mm), the overextrusion percentage may be substantially less.

What is perhaps more important to keep in mind from a design perspective is that the less your design relies on precision, the more likely it will be to work as intended without tweaking by the end user.

Test prints with known parameters are also very useful, as these parameters can be passed to the user as printing specifications to facilitate more faithful replication.

In addition to allowances for overextrusion, clearances must be allowed for parts to fit together. In any manufacturing process, A 10-mm peg *will not fit into* a 10-millimeter hole without deforming the peg, the hole, or both. Holes for free passing fasteners should be designed about 0.3 to 0.8-millimeters larger than the fastener diameter, in addition to any extrusion allowance. If a fastener is large and alignment is noncritical, an extra millimeter of radius is not excessive.

Screws, on the other hand, may be threaded into a hole 0.8 times their nominal diameter *after* overextrusion encroachment is allowed for. This rule of thumb applies to screws in the 2 to 6-millimeter range, but for more accurate information there are hole size reference tables in appendix III (page 220) for screws and holes that will be mechanically threaded.

For self-tapping or other specialty screws, manufacturer data should be consulted for accurate dimensions. If the material that the part will be printed with is hard, self-tapping screws should be used, holes tapped for machine screws, or the screw or bolt can be heated during insertion to form threads without over-stressing the part.

Large diameter threaded fasteners and holes can be printed, but have limited utility due to their comparatively large size and anisotropic properties. In general, printed threaded parts less than 10-mm diameter or with a thread pitch less than 1-mm are impractical except in special situations. Printed threaded holes down to 6-mm may meet with some success, but the need to finish threaded holes with a tap prior to use limits their practicality. Printed threads above 12/1.75-mm are a practical alternative to cutting threads, but may require some cleaning up with a tap anyway.

For glued or friction fits, joining pieces should usually be given a dimensional gap of 0.1-0.5-mm above any extrusion allowance between the interlocking parts to accommodate imperfections. This clearance may be omitted if post-print fitting is expected.

For snap or self-retaining fits, the parts must be engineered with sufficient flexibility to pass the interference point, but with sufficient rigidity and strength to resist accidental removal and provide the necessary integrity. This is often accomplished by providing a flexible surface on the part so that a portion of it may flex inward to pass the latch point. (see illustration of snap-latch assemblies on page 142) To create strong, reliable latching components, grain orientation and clearances will have to be carefully managed, being careful not to stress the parts in the Z-axis except in compression.

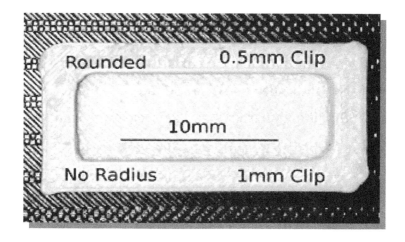

Photograph showing the effect of rounding (radiusing) or clipping model corners on resultant print accuracy. Abrupt changes in direction cause deceleration of the print head, resulting in over-extrusion due to the extrusion pressure lagging behind the change in travel speed. This effect is peculiar to the FFM process and varies by print speed, printer design and firmware. (Photo by author)

A printing-related issue that is useful to anticipate in the design process is that of overprinting at sharp corners. In the process of decelerating and re-accelerating in a new direction, the print head will often deposit a small surplus of material. This may result in a tiny spur at the apex of the turn. Spur artifacts can be reduced by using radiused or clipped corners. Slicers or printer firmware may reduce this artifacting by slowing in anticipation of the corner, or by anticipating the change in dynamic pressure of the extrusion plastic as the head changes speed.

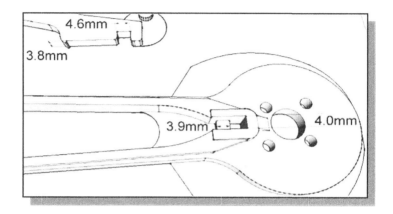

Quadcopter arm showing tolerances for fitting. The adhesion pad, or "brim", will be trimmed after printing. (Illustration by author)

Generally, corners should be rounded where possible. Even corners on tiny details can be clipped to good effect, as long as the X-Y length of the corner facet is roughly 0.5-millimeter or more. Rounded corners improve printing times, reduce artifacting, and decrease the chance of a motion related print failure.

CHAPTER VI: FASTENING AND JOINERY

Tying it all together

SCREWS AND BOLTS

The use of screws and bolts opens up many possibilities in functional design. Because they can significantly reinforce against layerwise delamination, threaded fasteners work especially well when oriented vertically on the Z-axis of the model. Multiple parts can be bolted or screwed together into a completed whole, and the result can be very strong.

Fasteners through the Z axis should be designed to load the part into which they are fastened in compression, as tension on the Z axis can easily cause part failure. This can be achieved by ensuring deep penetration into the engaged assembly, preferably completely through the part so that the fastener helps to "clamp" the part together.

When using screws in printed materials the screw must extend at least two times the screw diameter into a structure that is strong enough *in tension* to hold the anticipated load. When using screws, especially in the X-Y plane, care must be taken to ensure the hole is properly sized and reinforced to avoid splitting the part along its layers with the wedging forces from the screw.

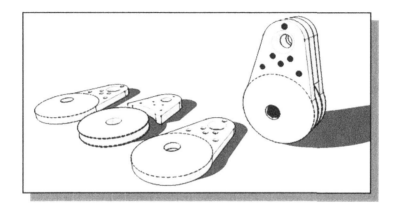

A few machine screws, an 8-mm bolt, 2 skateboard bearings, and a locknut are incorporated into this pulley design. Tested for working loads up to 100Kg when printed from Taulman 910 nylon, it would be much more complex to attempt this without a few inexpensive pieces of standard hardware. (Zombie apocalypse guide to 3D printing)

When in doubt about the strength of material holding an X-Y oriented screw, it is best to use through bolting or mechanically tapped holes with threads cut after printing instead of stressing a part with wedging forces.

Self-drilling, self-tapping screws also work well when screws are needed in an X-Y orientation.

A variety of special screws are available for use with plastic. For engineering guides and hole sizes for specialty fasteners, refer to the manufacturers provided documentation. For standard sizing information for generic screws (based on US sizes) and for bolts and machine screws in both metric and imperial dimensions, see Appendix III (page 220).

When assemblies are to be screwed, pinned, or bolted, it is often helpful to incorporate alignment bumps or tapered nesting elements in the design to ensure correct alignment and enhance rigidity. This helps to facilitate rapid and correct post-print assembly.

When designing pockets for bolts or screws, it is sometimes necessary for the part to be designed with a pocket for the head of the fastener at a lower (in Z -, so printed first) part of the print than the hole for the shaft of the fastener. This can be problematic to print without support, because the first layers of the smaller hole will be unsupported over the larger hole provided for the head.

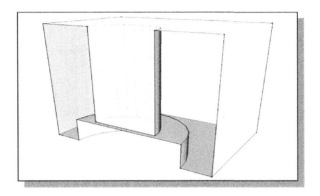

Cut-away example of a bridging membrane for a fastener hole. One or two layers is usually sufficient to ensure a properly printed fastener pocket. A thin membrane can be simply punched through when inserting the fastener.(Illustration by author)

This can cause very messy overhanging surfaces, and in rare cases can even lead to print failures. A simple remedy is to provide a thin (about 2 layer) membrane between the head pocket and the main body of the screw cavity. This provides a good bridging gap, creating a surface to support the first layers of the smaller diameter hole. The resulting membrane is then easily breached when inserting the fastener during assembly.

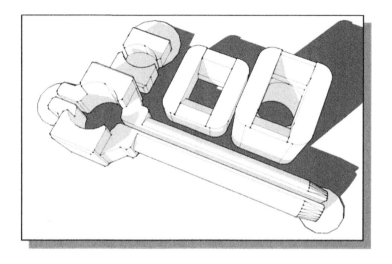

Split pin tubing joint model. The split pin halves are passed through the two collars (top center and right) to retain their placement and hold the head together where an intersecting tube (shown on the following page) will pass through the hole formed by the two halves. The small pieces in the upper left are the counterparts to the protrusions on the heads of the split pin, but are for the bottom (print bed) side. This assembly is designed to be printed in PLA and thermoformed, assembling the part after immersing in hot ($95°c$) water. This allows the parts to cool in a precision interference fit. (Illustration by author)

Split pin tubing joint, printed and partly installed. The long pin inserts into another metal tube which slides up into the collar, forming a reliable joint for use in an experimental drone platform. (Photo by author)

JOINERY

Dovetail joints, tabs and slots, printed pins and retaining collars can be used with excellent results when multiple parts need to be joined. With close attention to fitting dimensions, strong and easily assembled objects can be designed that do not require glue or fasteners.

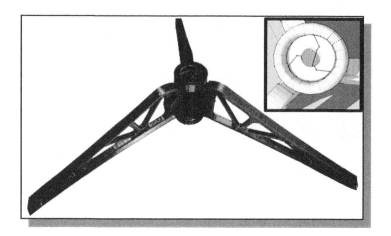

450-mm (18") Printed tripod base. Each of the legs are printed separately, flat on the print bed, and fit together in the center forming a tube. The tube is pressed into printed rings top and bottom. The assembly is rigid and strong, requiring no non-printed parts or glue. (The Zombie Apocalypse Guide to 3D Printing)

Collars can bear the outward pressing loads in an assembly with their circumferential shells, which are very strong in tension. They are an excellent choice when a group of parts can be encased in a load-bearing shell.

Dovetail joints provide excellent rigidity and precision alignment. If correctly sized, they will require a small amount of force to press together, and are self-retaining. This type of joint should usually be designed with about 0.1 – 0.5-mm of clearance, after allowing for overextrusion.

If carefully sized and used with locking features or adhesives, printed pins can largely negate the need for metal fasteners in systems that will not require disassembly.

Printed pins can align stacks of printed components and can be made to bear the tensile load in their ideal structural orientation using adhesives, locking tabs, or friction. Pins should usually be designed to fit tightly, so 0.05 - 0.2-mm of finished clearance (after allowing for overextrusion) is usually adequate.

Slots can often be used in cases where planes within an object will intersect. Think of two hands displaying the "V" gesture interlocking at 90 degrees. These interlock slot into slot and provide a stable, self-aligning fixture. Tab and slot interlocking plane systems are easy to design and provide fixation, alignment, and support with a high degree of design flexibility. They can incorporate "bumps" on the tabs and "wells" in the sockets to create a self-retaining snap fit system.

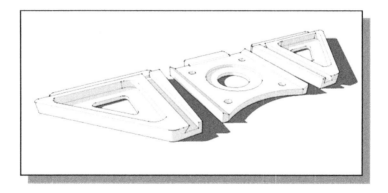

Dovetail joint design. 0.5-mm total free space was designed into this joint. Note that all parts lie flat during printing, maximizing strength and printability. (Illustration by author)

Careful use of joinery opens up a variety of possibilities that would otherwise not be accessible due to limitations of the FFM process, and can reduce weight, reduce the wasteful and time-consuming printing of support material, and greatly enhance functional utility.

The finished piece from the previous illustration, fully assembled. The grain of the parts is ideally oriented to provide strength to the finished assembly. Note the tabs at the bottom, forming a friction fit into the base (not shown). Once assembled into the base, the tabs will be glued in place. (Photo by author)

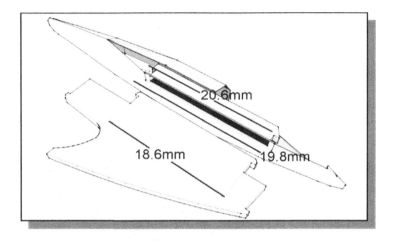

Cutaway view showing single use snap-fit details. Note that engineering tolerances allowed between the inserted part and the socket must accommodate for encroachment from overextrusion during the printing process. Usually, 0.4 to 1.0-mm is sufficient. An assembly such as the one shown could benefit from increased pawl area and relief in the tab to allow the pawls to be pushed inward during assembly. (threedsy.com)

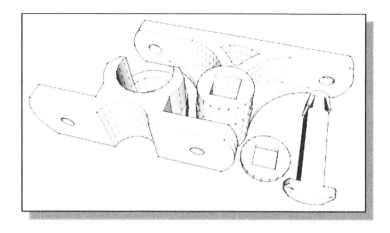

A half-kit for a pivoting flat panel display mount. Note the intentionally polygonal mating surface for a simple detent action, as well as the (easily printable) snap fit locking pin and thrust washer. In this design, two of the bracket kits shown above are used to make a complete upper and lower bracket set of four bracket halves, two pins and two washers. This breakdown simplifies the printing process and reduces material and time requirements over printing a larger two-part bracket set. A single, long pin aligning both kits might also be advantageous here. (Illustration by author)

CHAPTER VII: FUNCTIONAL ELEMENTS AND ENGINEERED FEATURES

Integrating function and features into your design

Functional design elements are often incorporated in an object to provide necessary features or facilitate post-printing assembly. Hinges, fasteners, and latching mechanisms are among the most common of these elements.

HINGES

Hinges can be of many types, but in all cases print anisotropy must be considered, and reliance on Z-axis strength minimized for good results. A common and very printable hinge variant is the flexible membrane "living hinge." Although many materials can successfully accommodate a membrane hinge, living hinge design is especially applicable to flexible materials such as nylon, PET, more flexible varieties of ABS and PLA, or TPU and other elastomeric filaments. In multimaterial design, Membrane hinges can be made from elastomeric filament and either printed with or later attached to more rigid assemblies printed from other materials.

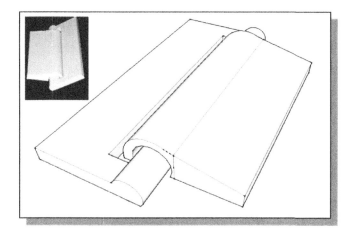

A semi-captive separated hinge design, removable when fully open (flat, as shown), captive when partially or fully closed. This hinge can be printed in place or separately. (Photo and Illustration by author)

Simple living hinges are most easily constructed by designing a thin 2 to 4 layer plane of 1 to 2-millimeters width, with bevels leading down to the thin section to accommodate the required range of motion without creating a leverage or interference point that would impinge upon hinge movement. Hinges can also be limited intentionally by the bevels and hinge width to create a fixed bend angle. This can be especially useful for hinges incorporated to aid post-print assembly. In this case, a surface suitable for gluing on the bevel of the hinge can help to achieve and retain the proper angle between parts for assembly.

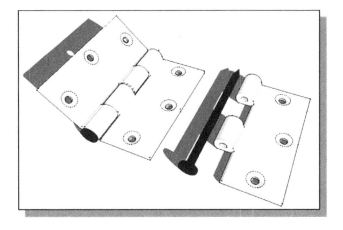

A conventional separated hinge design. Printed in hard nylon or PET, this hinge actually provides good service for lightweight interior doors. Interlayer adhesion with nylon and PET tends to be superior to many other materials, making this design practical. Note the use of a common 20d nail as a hinge pin. (Illustration by author)

When designing bevels and determining hinge width, it is helpful to bear in mind that printed perimeters are often slightly larger than their design dimensions due to nominal over-extrusion.

This is usually between 0.1 and 0.2-millimeters, so affording a margin of 0.3 or 0.4-millimeters should suffice to avoid binding in most cases. A tighter hinge can be utilized in elastomeric or very flexible filaments such as nylon, as they can be stretched repeatedly without weakening.

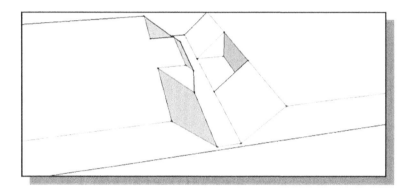

Living hinge with alignment tab. Living hinges can be printed out of flexible materials like nylon for repeated use, or more rigid materials when they will only be used in assembling an object into its final form, as in origami construction. They normally should be 0.3-mm – 1.0-mm thick, (minimum 2 layers, thicker with elastomeric filaments) with at least 1.5-mm of hinge width (bending space). For repeated use, they should be designed to be "loose" enough to have some slack when closed unless the material is very flexible. (Illustration by author)

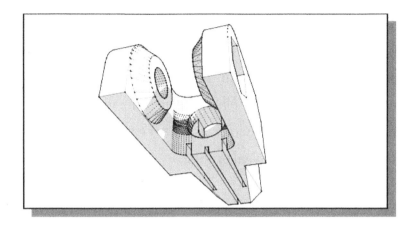

Idle roller fork and belt tensioner. Note the grooves on the underside of the part to improve tensile strength in the critical axis. (Illustration by author)

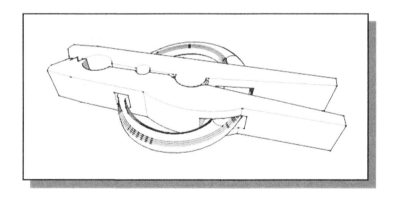

Printed clothespin. This design uses an integral printed spring, with minimal post-print assembly requirements. These have proven very effective and long lasting in PC-Max Polycarbonate filament, despite use in the intense tropical sun. (Illustration by author)

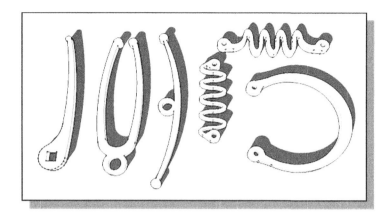

Printable spring designs. One disadvantage of incorporating parts like this is that the active qualities of the component will vary widely with printing process and slicing specifications, as well as with printing material choices. Printing springs requires careful specification by the designer and increased technical skill on the part of the printer operator. But, on the other hand... hey, it's a printed spring! Carbon fiber reinforced plastics can be excellent materials for spring construction. (Illustration by author - The Zombie Apocalypse Guide to 3D Printing)

LATCHING ASSEMBLIES

The possibilities for snap fitting assemblies are many, but careful attention must be given to the orientation of flexible parts and locking tab(s), with adequate flexibility designed in to accommodate the required deformation.

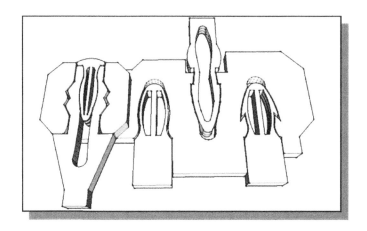

A variety of snap-locking designs, left to right: removable key pin lock (permanent if in a blind hole, or removable with access as shown), push-pull removable latch, snap in squeeze removable latch, and a permanent snap fit latch. Note the additional space provided in the sockets at the tip. (Illustration by author)

Additional space at the tip of the "socket" must be provided in most designs to allow a slight over-penetration, allowing the retention tabs to pass their catches. Snap latches can be made permanent or removable based on the form, flexibility, and accessibility of the interlocking parts.

Many other systems are possible through the evolution or combination of these primitives, and with thoughtful design, primary components incorporating latching assemblies may serve a dual purpose as fasteners - reducing complexity and assembly time.

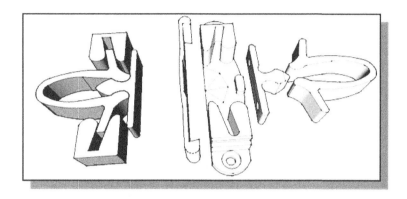

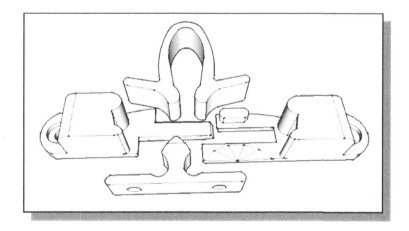

Two treatments of a self-releasing cabinet latch. Note the angle of the slots for the spring guides, helping to open the spring under closing pressure but tending to keep it closed from minor opening forces. (Illustration by author)

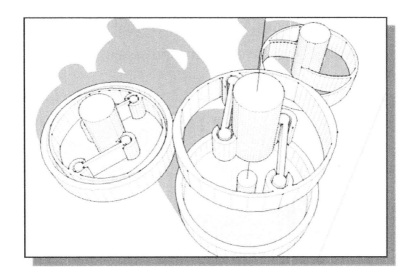

A light duty one-way drum clutch. Two varieties are shown. The clutch is made up of an outer drum, an inner drum, with actuating arms attached to the knob and the inner drum. Force is transmitted in only one rotational direction from the knob to the outer drum. The arms will distort the inner drum when the knob is rotated, forcing the lobes outward against the outer drum, causing the clutch bind tightly when turned counterclockwise. Note the thicker regions on the inner drum, 90 degrees from the arm attachment points. The inner drum is oval shaped, with the thicker lobes dragging lightly against the outer drum to provide the friction required to actuate the clutch. Gears, clutches, bearings, and other mechanisms can be printed at a very low marginal cost, enabling sophisticated mechanical functions within the limits of the material used. (Illustration by author)

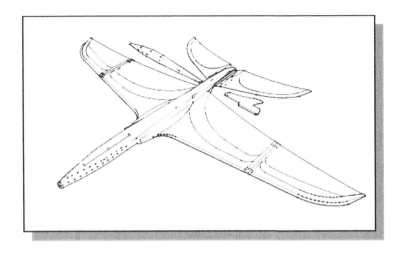

Advanced 3D printed flying glider, from threedsy.com. Handmade By Robots™ uses a combination of origami, design by slice, and conventional solid design to achieve performance equal to or better than traditional balsa gliders in an easy to print model. The polyhedral wings and V-tail fold into their assembled position and are secured by a drop of CA glue. Carefully designed stiffeners provide predictable flexion of the wings, allowing them to adapt automatically to changes in dynamic forces. This engineered feature allows them to unload the wing surfaces in high speed flight, providing better flying characteristics over a wide range of conditions. (threedsy.com)

MULTI-MATERIAL PRINTING

Dual or multi-filament printers enable a multitude of design possibilities. The most common use is for decorative or aesthetic use, and this offers startling advantages over the typical monochrome results from single material printers.

In some applications, the considerations are more than cosmetic. When designing living hinges, for example, a flexible material may be implemented in the hinge, enhancing strength, flexibility, and durability. A rubbery, flexible filament such as SemiFlex™, NinjaFlex™ or FilaFlex3D™ makes an excellent living hinge material, as does nylon. When coupled with rigid parts, this can open up entirely new design possibilities.

When designing multi-material parts of this nature, be sure to create an interlocking structure to ensure proper material bonding at the material transition seam.

A multi-material living hinge design. The dark material represents a flexible or rubber-like material. Note how the materials are physically interlocked, not reliant on adhesive properties which may fail after a short time. The uneven protrusion of the parts here is for clarity and would not be used in practice. (Illustration by author)

A straight seam between differing materials is susceptible to tearing failure if overstressed or after long use. Joints between materials that vary widely in their elasticity should be designed in such a way that they are trapped in place by structure and do not rely excessively on the often unreliable fusion between disparate materials.

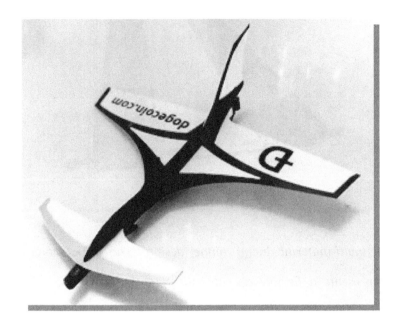

Dogecoin Flyer glider printed on a dual material printer. In this design, the color change happens uniformly at the third layer. Note that the same effect can be produced on a single color printer by changing filament colors mid-print. This and many other gliders designed by the author are available for download at threedsy.com. (photo courtesy of threedsy.com)

Sometimes it is advantageous to use flexible materials as inserts or bushings in a model to enhance the flexibility of the finished object, to create shock isolation, or to dampen vibrations. Flexible materials are also useful to sleeve holes where fasteners will pass through (especially screws), making them tight fitting without being impossible to assemble, when carefully executed.

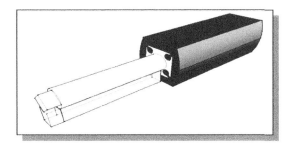

A printed, nonconductive trimmer adjuster with a rubberized grip. Note the support scaffold under the shaft and tip, designed to be printed in dedicated support material. (Illustration by author)

Flexible materials may also form the primary composition of an object. This is especially the case with objects that might traditionally be made of fabric. In these cases, it is often useful to incorporate hard features, such as the base on a small handbag, as well as the joints and closures where a flat-printed bag can be folded and joined into its final 3D form.

As with printing living hinges, it is important that the flexible material should have structural "roots" in the more rigid material, or vice versa, interlocking the two materials to ensure durability and long service life for the finished piece.

A multi-material living hinge using ABS and soft Nylon. Here the captive form of the flexible material is clearly visible. (Photo by author)

Rubber type filaments can also make ideal texturing or grip structures. Here, the bonding challenges are reduced, but even in this case, an effort should be made to see that each piece is entrapped by its form.

BUILDING LIGHT

Techniques that may be used to minimize weight, to reduce cost, or to increase flexibility include skeletonization and the incorporation of lightening holes. These techniques can be used to great advantage in creating lightweight, resilient structures.

When possible, such structures are best printed "flat" on the bed and assembled into their final configuration using interlocking, friction fit, or glued joints.

With careful attention to bridging widths and overhanging angles on the Z-axis, hollow beams and features can also be designed with excellent results. Keep in mind that any internal voids must be "vented" to the outside, or they may be ignored by your slicing software.

A method for minimizing material use is to think of your design in terms of tension in the X-Y plane. A part that focuses compressive loads at the center and perimeter, using tensile elements to do most of the work between the edges can be much lighter and stronger (by weight) than a part that is monolithically designed.

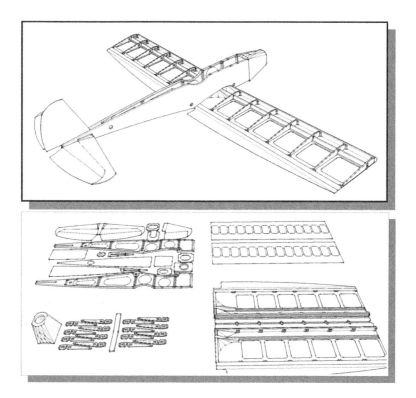

Careful design orientation and factoring can make it possible to construct lightweight, strong structures that print reliably with little or no support material. Here a flying rubber band powered airplane is shown, showcasing a 3D structure built almost entirely from membrane components. In this type of design, significant post printing assembly is obviously required. (Threedsy.com – Handmade by Robots)

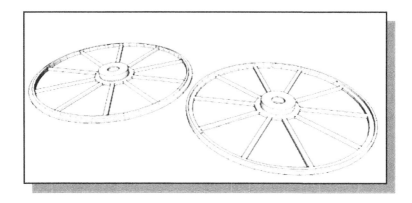

When designing parts with tensile components, it is sometimes useful to pre-stress the part to achieve the desired rigidity. Note how the wheel in the illustration above will be under stress when the hub is assembled.... The two wheel halves will be inserted into each other, tensioning the spokes and compressing the arc of the wheel rim as the two halves of the rim are joined. The spacing ring on the hub causes tensions the spokes as the hubs are pulled out-of-plane with the wheel rim during assembly, resulting in a rigid and durable part. (Illustration by author)

The classical engineering example of this is the spoked wheel vs. a solid one. Like the cables of a suspension bridge, the spokes act only in tension, while the circular arc of the wheel rim provides an ideal compressive structure acting against itself. The resulting wheel is both strong and light.

CHAPTER VIII: UNDERSTANDING THE SLICER

Design for the software

Even though model slicing is not strictly speaking part of the design process, many projects can benefit from the design finesse that comes only with a thorough understanding of the entire toolchain, from design to end product. Much like designing for other manufacturing materials and processes, knowledge about the actual method of construction is a useful (and sometimes critical) tool in the designer's toolbox.

With a thorough understanding of the process that converts a 3D model into the actual actions of the printer, it is possible to anticipate problematic situations and to minimize them in your designs. This can help you to avoid unpredictable outcomes when the actual software and hardware that will print your design cannot be known.

LAYERS AND SLICES

To be printed, a 3D model must be processed into a series of instructions to be executed by the printer. Like a series of 'maps', these instructions tell the printer where to move the print head, how much plastic to extrude, and how fast to move and change direction.

When the printer completes one layer of printing, the print head is moved up a fraction of a millimeter (the slice thickness) and the next layer is printed according to the next 'page' in the 'map'.

This stack of 'maps' that make up the instructions for printing the model are made by *slicing* the model into a stack of 2-dimensional layers in the Z-axis. This is typically done by a piece of software called a slicer. Slicers for 3D printing are available in free and open source varieties, as well as commercially.

Proprietary, commercial slicing tools can sometimes offer more automation, control, and flexibility to the operator, but the specifics of these tools are outside the scope of this book – though the principles herein generally apply to those tools as well.

With common slicing tools, the model is processed into its layers using a few fundamental parameters. The most universal of these parameters are layer height or slice thickness, the number of perimeters or shell thickness in the X-Y plane, the top and bottom surface thickness, and the infill percentage or density. Sometimes also available and relevant to design are infill orientation and infill geometry adjustments.

Layer (slice) thickness is the height of each printed layer. In general, most printers today are capable of 0.1-millimeter (100 microns) slice heights or thinner, up to around 0.3-millimeter. A thinner slice height will yield smoother vertical surfaces in the Z-axis and better closure of overhanging structures, as the overhang per slice will be a smaller proportion of the nozzle width, resulting in less unsupported material. Thinner slices can also result in lighter prints by making thinner solid layers at the top and bottom, but at a proportional cost in printing time (and by extension, overall process reliability.)

SHELLS: PERIMETERS AND SOLID LAYERS

The number of solid layers dictates the solid layers at the Z terminus, or top and bottom, of any models or printed features. Two solid layers are usually considered a minimum, as less will often result in the incomplete closure of upper facing surfaces or a weak, permeable face. Increasing this parameter can give prints more strength; but at the cost of increased material use, increased printing time, and sometimes problems with heat buildup on converging gently sloped surfaces. When very low (<20%) infill settings will be used, it may be necessary to increase the number of solid layers to get acceptable top surface closure and finish, as the first "solid" top layer may sag into the infill significantly. Localized cooling helps to reduce this problem. For non-structural purposes, a top and bottom layer thickness of 2 layers or at least 0.3-millimeters is usually sufficient.

The number of perimeters specified dictates the number of contiguous solid shells that each layer will have at its outer or inner edges, determining the vertical wall thickness of printed features. If the engineering requirements of the design are not too technically demanding, two perimeters is usually sufficient and even one perimeter can give satisfactory results in simple, lightly loaded structures.

When there is a requirement to carefully manage shell construction for strength, weight, or other reasons, extra care must be taken with the number of perimeters specified and the thickness of critical wall structures.

Where solid walls or shells are desired, the number of perimeters times 2 (for inner and outer surface) multiplied by the nozzle thickness should correspond to the wall thickness of the structure to ensure solidity and to reduce the possibility of a poorly bonded structure.

Example:

> *Desired solid wall thickness: 2mm*
> *Nozzle width (N) 0.5mm*
> *Number of perimeters (P) to specify: 2*
> *Because: (2P)N = 2mm*

A small deviation here can result in thinner than intended walls, or may create gaps between wall surfaces.

Where solid walls are not desired but completely hollow walls are not acceptable and must be infilled, some slicing software requires the space between the specified number of perimeters to be at least 4 nozzle widths. This figure depends heavily on the infill algorithm and printer, but 4 nozzle widths is usually reliable for triggering the infill algorithm and ensuring reliable infill construction.

For example, for an infilled wall of 3-millimeters with a 0.35-millimeter nozzle width, 2 perimeters will yield a total of 1.4-millimeters of solid shell with a gap of 1.6-millimeters between. A gap of 1.6-millimeters should be ample to trigger infilling, even if the specified infill percentage is low. A single shell will also work and would permit wall thicknesses as fine as 2.1-millimeters without risking a completely hollow core or poorly bonded infill.

Example:

> *Desired wall thickness (W): 3mm*
>
> *Nozzle width (N) 0.35mm*
>
> *Maximum number of perimeters (P) to specify: 2*
>
> *Because: $W-((2P)N)<=4N$*

poorly bonded infill (bottom) due to a wall specification that was poorly specified for the slicing software. A thinner overall wall fills in nicely (top), while a thicker wall than shown would have properly printed infill, using the same slicing software and settings - Cura, in this case. (Photo by author)

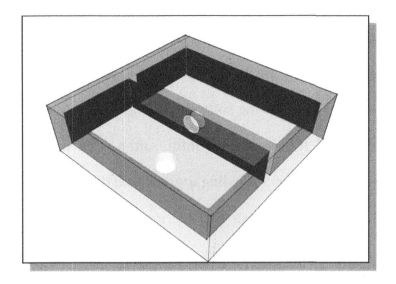

Venting of internal voids to improve slicer recognition. Note that at least one void must be vented to the outside, while others may be vented through the vented void or directly to the outside. If voids are not vented to the outside surface of the print, many slicing algorithms will fail to interpret the void walls as being valid surfaces, ignoring them. (Illustration by author)

Where voids or hollows are desired in a piece, **the internal void area must be connected or 'vented' either directly or indirectly to the outside of the piece**, as most slicing software will ignore "orphaned" voids and fill them with infill. Often, this is easily accomplished by providing a small hole down through the Z-axis to the base of the print, but any means to provide a vent should work.

By understanding the slicer and anticipating its actions in the execution of the printing process, the designer can create models that print more reliably and will print true to the intentions and engineered requirements specified in the design.

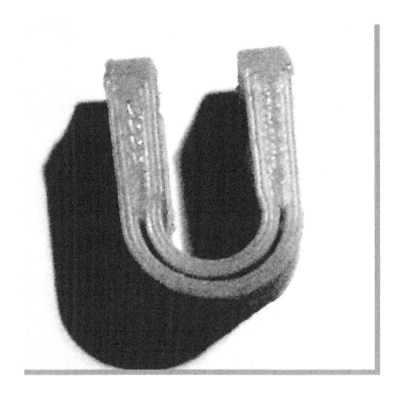

Bar clamp for a Mendel Prusa Printer. Note how the two walls of the curved portion are not connected. This is because the wall was specified too thin to accommodate infill with the 3-perimeter wall thickness, but too thick to evenly match 6 perimeters without a gap. Even though it doesn't appear in the model in this case, the gap was intentionally made by the printer operator, enhancing the flexibility of the clamp. Where the wall is slightly thicker, the slicer infilled between the perimeters, creating the solid sections shown above. In practice, it would of course be better to design the gap into the model instead of relying on operator finesse. (Illustration by author)

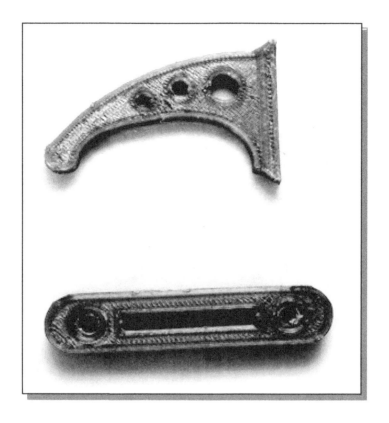

A two piece wall hook. Note the structure of the upper piece: The holes are reinforced with multiple perimeters and extend down through the otherwise unreinforced area of the part, improving interlayer adhesion and rigidity. The reinforcing holes require the specification of at least two perimeters, preferably 3 or 4, to be effective. (Photo by author)

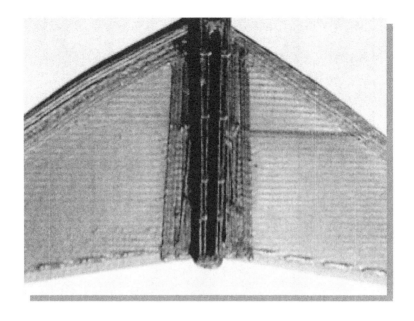

Note the thin gap in the walls to the left and right of the rudder socket on this toy glider. A single shell and low infill setting will cause this on very thin, vertical walls that are not exact multiples of the nozzle width. Here, reinforcing "stitching" is provided by sloping the top of the print in Z, causing the layers to terminate at different parts. Sometimes, the designer can anticipate what the slicer might to do to her models before printing and implement an 'outside the box' solution such as this. The error (line) visible on the right panel is a mechanical printer error, caused by excessive belt looseness. (threedsy.com)

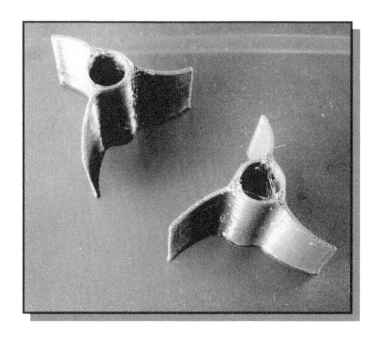

With careful control of wall thicknesses and nozzle width settings, models with critical features can be made to print reliably on most software stacks and printers. These arrow fletchings were specified with a 4-mm nozzle width, 2 perimeters, and 0% infill, with the minimum feature width in the model being 0.75mm. Note the transition area from the fins to the tube, reducing stress concentration in this critical part. (Photo by author)

APPENDICES

APPENDIX I: MATERIALS FOR FFM PRINTING

ABS, or acrylonitrile butadiene styrene, is popular due to its low cost, pervasive use, and excellent mechanical characteristics. It prints well at easily achievable extrusion temperatures, and can be very strong and rigid.

The ratio of different components - as well as the dyes used for color - determines the character of ABS filament. A large styrene component, for example, gives increased rigidity and brittleness, whereas increasing the butyl fraction makes it more flexible. Some dyes can materially raise or lower the glass transition floor of the plastic, necessitating different temperature settings than another otherwise similar composition would require.

ABS can be glued effectively with cyanoacrylates, specialized epoxies such as JB Weld™ PlasticWeld or Loctite™ Epoxy Plastic Bonder and solvent-welding cements commonly used for plumbing. This makes ABS an ideal choice where adhesives will be needed in the final assembly process, greatly enhancing design versatility at the expense of complex post-print processing.

ABS has a rather high glass transition floor and a high thermal shrinkage dynamic, making warping and bed detachment a challenge for some prints. Various methods for overcoming these difficulties are described later in this text.

Because of its multi-component nature, a large variability exists within ABS filaments, but for the most part ABS is used when strength and rigidity are required with a non-exotic filament composition.

ABS filament is widely available from a myriad of sources in a variety of colors, diameters and product qualities.

PLA, or polylactic acid, is a bioplastic derived from corn or other dextrose sources. It is biodegradable, transparent and well-suited to 3D printing due to its comparatively low glass transition and extrusion temperatures. It can be thermoformed after printing by dipping in hot water, and tends to return to its printed shape upon reheating. It has much less warping and thermal distortion than ABS and ranges from very hard and somewhat brittle to rubbery, depending upon the composition and the proportion of additives. Because of its natural transparency, translucent colors are widely available.

PLA filament is widely and inexpensively available in a variety of colors, diameters and compositions for use in 3D printing.

Because of its low extrusion temperatures, low shrinkage, good mechanical characteristics and wide availability, PLA is often the first choice for entry level printing, and some printers cannot print other plastics without modification.

One of the principal drawbacks of PLA plastics is the relatively narrow choice of options for using adhesives with the material. If the part will not be subjected to high loads, CA glue may be sufficient. Choices for structural bonding include welding, hot-melt adhesives, 7515a11 Scigrip Plastic Pipe Cement from McMaster-Carr™, Bondene from Plastruct™, and a few others. PLA adhesion is well below material strength with CA glues, and epoxies offer a similarly weak mechanical bond.

HIPS (high impact polystyrene) is similar to ABS, but generally slightly more brittle. This is the stuff that glue together plastic model kits and CD jewel cases are made of. A styrene alloy with polybutadiene as a plasticizer instead of acrylonitrile butadiene, its printing qualities are excellent and closely mirror ABS. HIPS features lighter weight, less warpage but less impact resistance than ABS. It can be dissolved in a limonene solution, so HIPS can also be used as a support material for ABS which is unaffected by limonene based solvents.

HIPS filament is widely available in a variety of colors, and prints well on any machine that prints ABS. HIPS can be solvent-welded with limonene based or other "model glues", as well as cyanoacrylates, epoxies, hot glues, and most other adhesives.

PET / PETG / PETT (polyethylene terephthalate) is a very durable, extremely versatile plastic that is used extensively in the manufacture of plastic beverage bottles, textiles (Dacron), as well as impossible to open plastic packaging. Somewhat more difficult to print than ABS (a lot of stringing issues) PET gives very durable prints, and can be obtained in opaque, transparent and translucent colors.

Widely popularized by Taulman under their trade name Tglase, PET filament is now widely available from a variety of manufacturers. One difficulty in modeling with PET is that it can be difficult to glue. Among a very few adhesives ideal for PET, acrylate adhesives such as Cyberbond Cyberlite U320 or Dymax 3081 are commonly used UV and visible light curing resins. Mechanical bonds with elastomeric glues are also possible, but do not approach material strength. Hot melt glues and flexible silicone or acrylic adhesives can be used with variable efficacy depending upon the application.

PC (polycarbonate) is a very durable printing material, but requires a higher extrusion temperature than many printers can safely manage. Most PC filaments print in 290-310° Celsius range, but a growing number of manufacturers are producing lower temperature offerings. Polymaker's PC-MAX is a good example, printing very nicely at 260° (250-270°).

PC-Max is an exceptionally strong and easy to print variety of polycarbonate, and adheres well to a prepped Kapton or PEI surface with less warping than a lot of ABS filaments. Polycarbonate tends to be extremely hygroscopic, so drying and keeping the filament free of absorbed moisture is critical in order to avoid steam bubbles (foaming) in the extrusion process.

Despite the technical hurdles, PC is unsurpassed in durability, and is the material of choice for bulletproof glass and aircraft windows. It is available from a small but growing list of manufacturers at this time. Polycarbonate is somewhat unique in its ability to be cold-formed without failure, and it feeds reliably in hobbed-bolt and direct-drive extruders.

Polycarbonate can be bonded with solvent welding (chloroform or methylene chloride), some CA glues such as Cyberbond™ Apollo 1603 and specialized epoxies such as JB Weld™ PlasticWeld or Loctite™ Epoxy Plastic Bonder.

NYLON is an extremely tough, flexible plastic that can be printed by many off-the-shelf machines. It adheres to cellulose-based printing surfaces (paper, wood and others), but performs best on a garolite bed. Some newer nylon alloys, such as Taulman bridge nylon, adhere satisfactorily to a PVA (Elmer's Glue-All) treated surface and have much less shrinkage and curling than pure nylons. Taulman's 910 "alloy" filament is an excellent, high strength alternative to many other nylons and has excellent printing characteristics for a nylon.

Nylon is very hygroscopic, so drying and managing moisture is a significant issue. Taulman "bridge" nylon significantly moderates this problem, so that little drying is required compared to other nylons.

In practice, nylon can yield excellent results where tensile strength and flexibility are important. Its extrusion temperature is at the high end for most machines at 245° Celsius, but most PEEK insulated hot ends will handle this temperature without difficulties.

Nylon can be bonded with CA glues and epoxies such as JB Weld™ PlasticWeld or Loctite™ Epoxy Plastic Bonder.

PVA, or polyvinyl alcohol, is soluble in water and makes an excellent choice for a support material that can be used in most FFM printers. It is not normally considered for use as a primary construction material.

Thermoplastic Elastomers (TPE/TPU/TPC)- flexible filaments such as NinjaFlex™ or FilaFlex3D™ vary in composition, but TPU (Thermoplastic Polyurethane) is the most common base stock. Most require special extruder drives to feed reliably, but NinjaFlex™ Semiflex and others maintain enough rigidity to give good results with most printers.

Acrylic (PMMA) (Polymethyl methacrylate) is a rigid, impact resistant material normally selected for its transparency. PMMA can be solvent welded with acetone and bonded with acetone based glues.

Acetal (POM) Polyoxymethylene, or polyacetal, is a low friction engineering plastic with high rigidity and impact resistance. It is a strong plastic that prints at relatively low temperatures and is normally selected for applications where its rigidity and low friction characteristics are required, such as long life gears or bushings.

Acrylonitrile Styrene Acrylate (ASA) is a material with great promise for long-lasting 3d printed objects. There are few filament vendors as of writing. The material tends to be brittle, but it has a hard, weather-resistant character that makes it ideal for many types of functional objects. It can be glued with CA glues and Acetone based solvent adhesives.

Polypropylene is extremely chemically resistant, highly flexible and very durable. Almost impossible to glue, this material is food safe and a good choice for items that may take a lot of deformation abuse. Filament choices are few as of writing, and polypropylene is difficult to print, being similar to pure nylon in many characteristics.

"Alloy" filaments

Most common among specialty filaments are the composites, or "alloys," which combine powders of wood, stainless steel, iron, bronze, sandstone, chopped carbon fiber, graphene, carbon nanotubes, or a host of other materials with PLA, ABS, nylon, PETG, or other base thermoplastics. These give the printed object some of the characteristics of the supplemental material such as finish, magnetic ability, strength, rigidity, conductivity, or texture.

New materials are being introduced all the time, and a wide variety of filaments with unique properties are available from a growing variety of sources.

Filaments with special properties like Multi3D's highly conductive polyester and copper *"Electrifi"* are beginning to make printable electronics a reality, while Taulman optical Tglase and Guideline filaments make printing light guides, bezels and transparent or translucent objects simple and reliable.

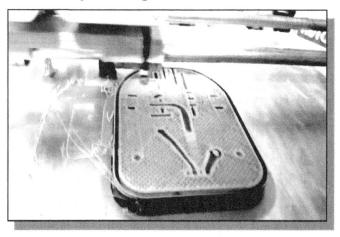

Multilayer 3d Printed electronic circuit board for a microcontroller-based platform. By leveraging new materials and multimaterial printing, exciting new frontiers in desktop manufacturing are being explored. (Photo by author)

Characteristics like thermal or moisture based color change, sponge-like absorption, luminescence and other material characteristics are widely available.

Carbon fiber reinforced materials are quickly becoming a revolutionary tool in the FFM designers toolkit. Carbon fiber alloys using nylon, PETG, and other base stocks combine metal like stiffness with the toughness and light weight of modern thermoplastics.

By the time this goes to press, there will probably be tens or hundreds of new filaments on the market in this fast growing field, and each new offering brings the promise of expanded functionality.

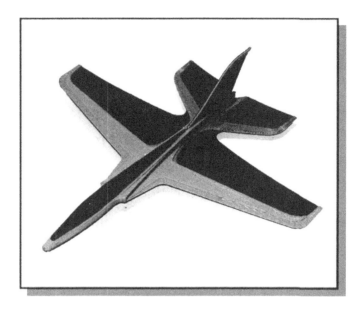

Multicolor 3d Printed glider. Leveraging the extreme durability of filaments like PC-MAX, this toy glider performs as good or better than similar gliders made of more conventional materials such as foam or balsa, and is nearly indestructible in normal use. (Threedsy.com – Handmade by robots ™)

APPENDIX IIA: PRINTING ELECTRICAL CIRCUITS USING FUSED FILAMENT MANUFACTURING (FFM)

The rapid design and manufacture of complete functional assemblies is one of the great promises of 3D printing. To this end, a great deal of research and development has been invested into developing material systems for printable electronic circuits using the FFM process. Unfortunately, the results up until now have been a bit disappointing, with high resistances, poor connection durability, and suboptimal printability in standard printers.

Fortunately, some significant developments are emerging that could open the door to fully printable circuit boards and even some printable passive components.

Most recently, I have had a chance to try out some filament from Multi3d, which shows great promise and could represent a crucial first step in the emergence of FFM printable electronic devices.

Multi3D's (www.multi3dllc.com) polyester and copper *"Electrifi"* filament features a comparatively low resistance at between 0.01 and 0.09 ohm/cm. At the time of writing, this represents the state of the art, turning in resistance figures about 50x lower than its nearest competitor.

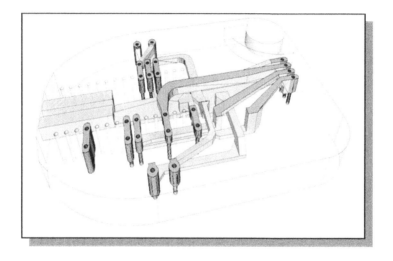

Model of an all printable Arduino computing platform with an OLED display, two menu buttons, and an analog input. Note the three layer design made trivially printable through multimaterial printing. (Illustration by author)

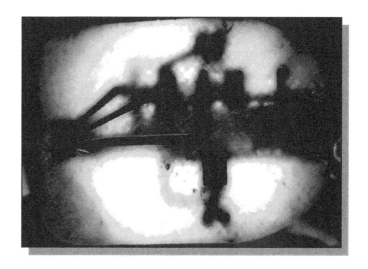

The printable Arduino computing platform shown earlier, printed. This photo is backlit and contrast enhanced to show the various layers in the circuit board. (Photo by author)

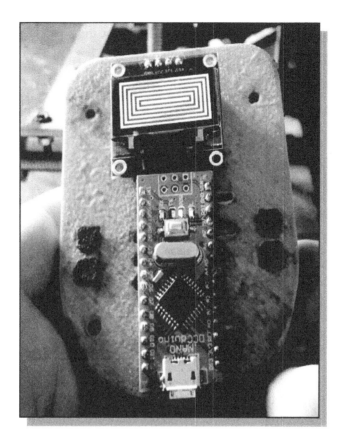

The Electrifi printed Arduino computing platform in operation. Printed switch contact pads are located on the left and right of the CPU module. Note the cosmetic contamination of the board caused by the wide disparity of melting temperature between the Electrifi filament and the ABS that was used. This would print cleaner and smoother using a lower temperature substrate. (Photo by author)

At these levels of conductivity, conductive filaments can be used to carry power to light higher power LED's and to power microprocessors and other common components. It even has the ability to form a solder – like bond with components, although the use of conductive adhesive is still needed in many cases.

The model shown was printed using Electrifi and ABS, and while fully functional, required the use of conductive glue for the contacts to achieve reliable operation.

I am excited to see further development along these lines and will continue my experimentation with *Electrifi* and upcoming conductive filament systems as the technology matures.

Although *Electrifi* conductive filament represents a huge step forward for FFM printable electronics, conductive filaments have yet to achieve the utility necessary to replace conventional manufacturing processes in most applications. In particular, the three major milestones necessary to achieve a utility level of performance from this technology would be:

1. **Pluggable component and wire connection.**
So far, existing filaments cannot form reliable plug-in connections that would allow the trivial insertion of IC's, resistors, connectors, or other components into the printed assembly. This capability is key to replacing existing technologies – if you have to use a conductive glue or a heat process to ensure a good connection, it's usually actually easier and more reliable to just "draw" your circuit using conductive cement in printed channels on the circuit board, or even on a flat insulating substrate.

2. **long-term stability for durable projects.**
Existing solutions quickly fail in humid or corrosive environments and have a limited lifespan even under relatively benign conditions unless special treatments are used at the connections. Printed electronics will have to withstand long-term real world exposure in order to transcend the experimental stage.

3. **Low cost and ease of implementation.**
   Existing printable conductors are more expensive than other solutions and often require special termination. This is fine for the experimental stage of development, but the price and effort barrier will have to come down in order for the technology to see significant adoption.

Considering these requirements, I spent some time developing a system of printable "circuit boards" for electronic projects. While these do not achieve the goal of "print and play" by any means, they provide an acceptable interim solution for the easy implementation of low power circuits using the FFM process.

The systems shown consist of "wire wrap" and "breadboard" types and differ in their method of construction.

The breadboard type uses clips from a standard solderless breadboard, and essentially consists of printing a project-customized breadboard that can then be trivially wired up and used. This has the advantage of resulting in a wholly usable project.

Traditional breadboards strictly dictate component placement, which makes fully usable prototypes impractical to achieve in most cases.

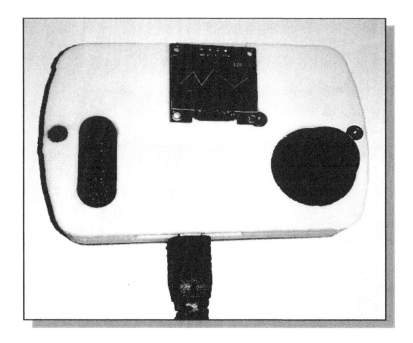

A printed Arduino / NodeMCU based development platform. SPI and I2C displays are supported; shown here with an 0.96" OLED, analog joystick and 2 menu buttons. (Photo by author)

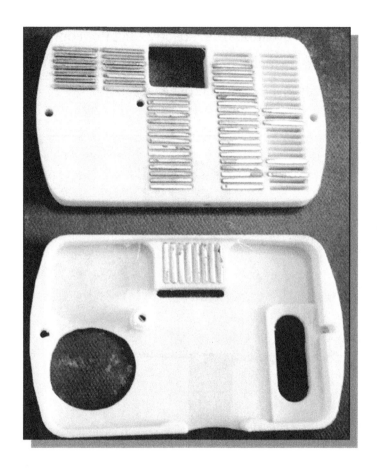

Detail of breadboard clip placement in the printed development platform. Clips from a standard solderless breadboard are inserted from the rear. (Photo by author)

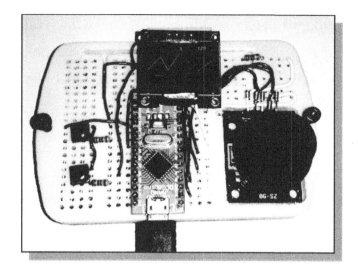

Populated circuit, showing component placement and wiring. The design allows wiring variations and the incorporation of additional components in the small prototyping area. (Photo by author)

The "wire wrap" printed prototyping system is a printed armature around which wire wrap wire is wrapped and routed. This differs from traditional wire-wrap circuit building in that the component leads themselves are not wrapped, but rather the plastic armature.

Wiring the armature is a bit tedious, but the result is more compact, durable and customizable than the "breadboard" style. Additionally, the wire wrapped armature provides functional sockets for components and provides a great deal of design flexibility for component and trace location.

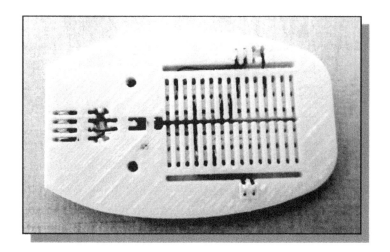

Wired armature, demonstrating wire pathways and routing. Component sockets are made by stripping the wire ends and wrapping the armature with the stripped wire where contact will be made. The holes are slightly undersized, providing a tight reliable fit. Prototypes show excellent ruggedness, reliability and resistance to degradation – even after over a year in a humid tropical environment. Note the wrapped, elevated switch contacts on each side of the main component area. (Photo by author)

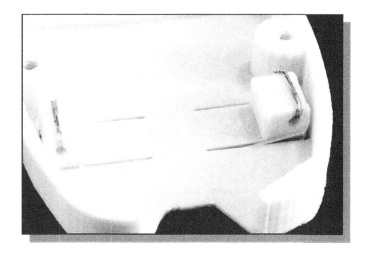

Detail of button contacts. When the buttons are pressed, the wire shown will connect the contacts on the main board in the previous photo. (Photo by author)

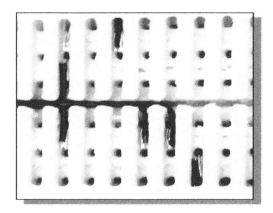

Detail of wire wrapping. Each wrapped hole becomes a junction, a pin socket, or both. Channels are designed in to allow easy routing of wires from one contact point to another. (Photo by author)

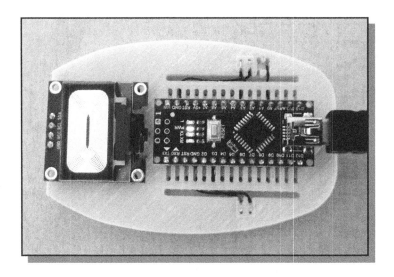

Populated wire wrapped board. Note exposed switch contacts (Photo by author)

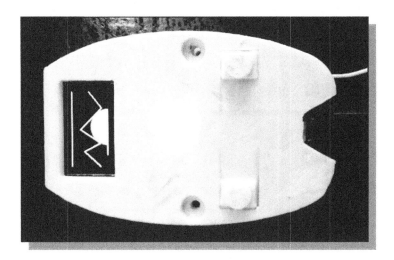

Wire wrapped board, populated and covered. Note the simple buttons integrated into the cover. (Photo by author)

Printed buttons and switches are easily achieved using the wrapped armature method. By using the wrapped wire as contacts and contactors, switch assemblies from simple pushbuttons or slide switches to custom rotary selectors can be printed.

When designing an armature for sliding switches, it is important to incorporate a spring mechanism in the design to maintain pressure on the contacts. This will allow the contactor to ride over the contact surfaces while maintaining even pressure for smooth and reliable operation.

Detents to hold the contactor on station are also desirable in many applications. These can be simple on-off bumps at the junction of sliding parts or multiple detents for a rotary or linear selector.

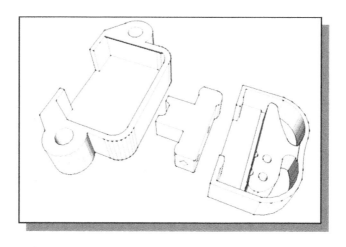

Sliding SPST switch. Note the small detent bumps at the end of the slider (center) and on the inner surface of the inner housing (right). (Illustration by author)

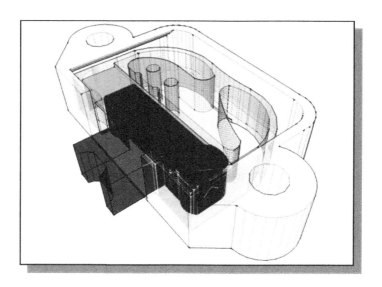

Assembled switch from the previous illustration. (Illustration by author)

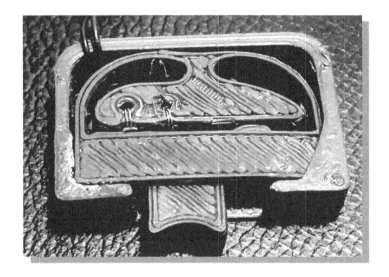

The printed switch modeled in the previous illustrations, printed 15-mm wide. Note the wire wrapped through the holes and the spring holding the contact assembly against the slider. (Photo by author)

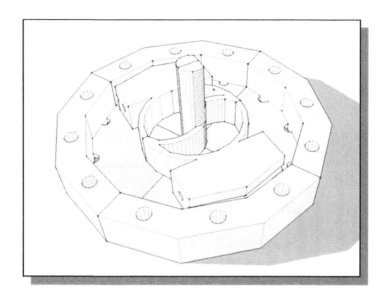

Armature for a rotary selector switch with the cover removed to show detail. This armature would be wound with bare wire through each of the twelve holes on the housing and around one end (or both ends) of the rotor. It will connect adjacent contacts as it is rotated through its positions. (Illustration by author)

The illustrations on the following pages *show additional detail of the two microcontroller development systems, with a printed battery holder tray included for the breadboard clip based version in the second illustration. (Illustrations by author)*

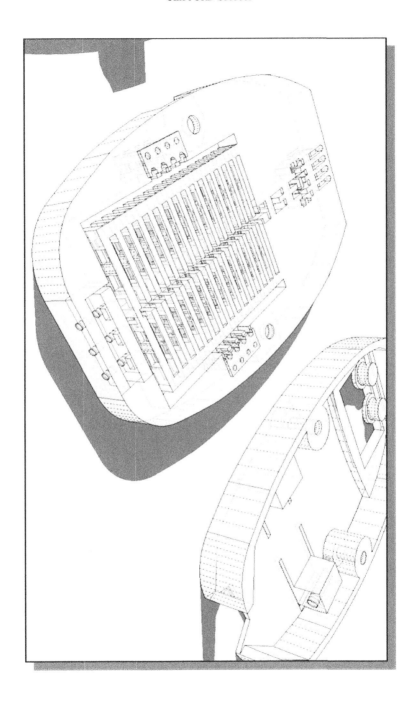

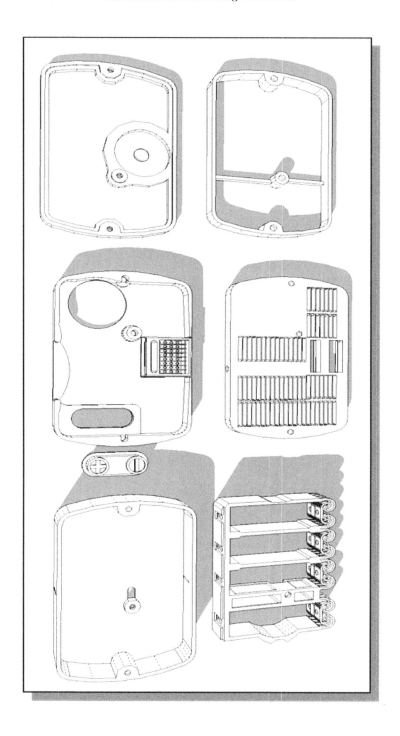

APPENDIX IIB: ANISOTROPY IN FFM PRINTING

While seemingly magical in its ability to transform ideas into physical form, 3D printing has its strengths and weaknesses – just like any manufacturing process. These limitations come in the form of constraints on the materials we can print in, the specifics of the printing process, and in the physical properties of the finished part, as determined by the interaction of these factors.

Parts built on an FFM printer are stronger in some directions than others. The directional strength, or anisotropic properties, of objects made using the FFM process superficially resembles the anisotropy, or grain, of wood. It is similar to wood in that there is an axis along which it is weakest, and may break more easily, especially in tension. It is different than wood in that this weakness has only one axis, rather than two.

Some of this similarity results for the fibrous construction of both materials. Wood is fibrous due to the nature of the cellulose structures made by its growth, and FFM objects are fibrous by virtue of the extrusion process, which extrudes a tiny filament of plastic which is welded together with the other fibers alongside it.

Photo showing a printed sample and a wooden twig split along the direction of their "grain". (Photo by author)

Most wood is strong in compression in all directions, but especially so with the 'grain' of the wood, or along the fibers (not the rings) of the structure made by the growth process. Similarly, wood is very strong along the axis of its fibers in tension - so remarkably strong in fact, that it rivals modern exotic composites.

Objects made on an FFM printer have similar properties, but with some important differences. In Compression, FFM parts tend to be strongest along the Z, or layer-wise axis, going up and down as the part sits on the print bed, though this can vary significantly based on finished shape and its internal structure.

In contrast to wood, the up-down, or Z-axis is *also* the weakest when it comes to tension. FFM parts tend to pull apart most easily along the Z or layer wise axis, with the weak point being incomplete adhesion between layers and microscopic variations in wall thickness or alignment.

Another important difference between FFM and wooden objects is - at least for simple solids - that there is only one axis of tensile weakness (Z, or up and down), with two axes of tensile strength (X and Y, or side to side and front to back on the print bed), instead of the single axis of strength found in wood oriented along the direction of its fibers.

As a designer, it is useful to remember that since the layout of the fibers in each X-Y layer is determined largely by the shape of the object being printed, the designer can easily influence the fiber orientation and the finished strength along these axes by varying the design or printing orientation of the printed piece.

Because of the directional, or anisotropic, strength properties of 3D printed objects, it is often useful to break down compound parts into components rather than print them in one piece, in order to make them as strong or easy to print as possible.

As an example, I give you the following practical object, a tarp clip which allows the user to attach a rope to any point on a piece of fabric without making a hole.

It is used by putting the 'ring' on one side of the fabric, then pushing the 'lock' through from the other side, turning it ninety degrees so that it locks into place.

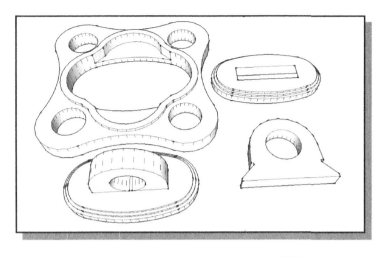

Tarp clip design used for testing. Note the two different styles of lock. (Illustration by author)

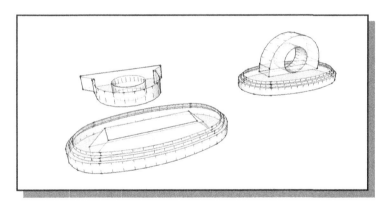

Structural detail of the two locks. (Illustration by author)

In the previous drawings, you can see that there are actually -two- locks represented....one printed in one piece, and the other printed in two interlocking pieces. We will explore the merits of both approaches.

The 'lock' part of the tarp clip incorporates an eye, to which a rope will be attached. This line will subject the eye to tensile loads, and we will test the resulting strength of the assembled part.

After printing, the compound lock must be assembled by inserting the eye through the base. This enables the eye to be printed in an orientation where its fibers will form a strong loop, and its wedge shape will engage the base in such a way that the tension on the eye will be a *compression* load on the structure of the base. In contrast, the one piece lock requires no assembly, but all loads will be on the layers in tension. This will influence the strength of the part.

For this test, we will attach the tarp clip to a piece of fabric as designed. We will then suspend a weight from the lock eye, increasing the load up to the point of failure. We will then examine the results and draw any useful conclusions from the experiment.

With the one piece lock, the bucket was filled to about 4.5 gallons of water (about 36 lbs.), when I started to notice sounds of impending failure. It failed shortly thereafter with no additional weight.

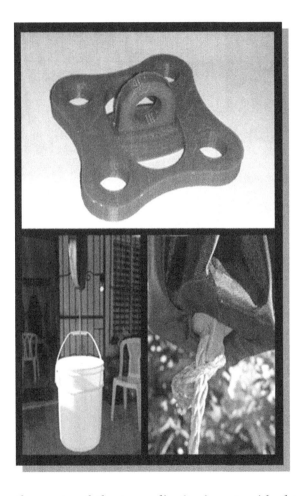

In use, the parts of the tarp clip (top) nest with the fabric between. Also shown is the bucket used for testing weight, and the tarp clip attach point. (Photos by author)

The failed one piece lock. Note the failure of interlayer adhesion in the infill, and the slight heat deformation of the attachment eye. (Photo by author)

After the 1 piece lock failed, I re-rigged the test for the compound lock. I filled up the bucket the rest of the way (Approximately 40lbs overall) but the lock did not fail. I then put a board on top of the full bucket, and set an approximately 30 lb rock on it. Unfortunately, as I was getting ready to take more pictures, the ring inside the fabric failed.

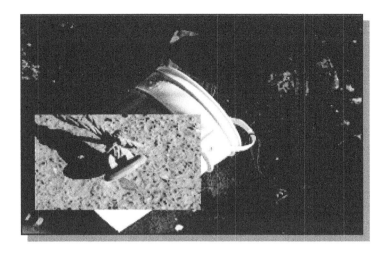

The scene after the ring failed. Note the large rock and that that the locking part is fully intact in this case. (Photo by author)

On examination, it became evident that the lock had pulled the outer ring apart, leading to an early termination of the test. The lock was undamaged and showed no whitening or other signs of stress.

Based on earlier tests making printed plastic carabiners, I estimate that the lock alone would have supported at least an additional 30 lbs. above the 70 lbs. that caused the ring failure.

Examination of the failed one piece lock showed that it failed in tension due to layer separation where the eye retainer merged to the base.

With the one piece eye, I was surprised that the eye itself did not fail as I had expected it would. I believe that this is probably due to superior interlayer fusion in the eye, caused by excessive heat buildup during that part of the print. You can see some deformation in the close-ups caused by the lack of adequate cooling, though in this case, I suspect that the excess heat vastly improved the performance of the part.

This experiment highlights the influence of the designer in determining the final strength and utility of printed objects. Here, two nearly identical pieces were printed in different forms and print orientations, with widely divergent strength outcomes.

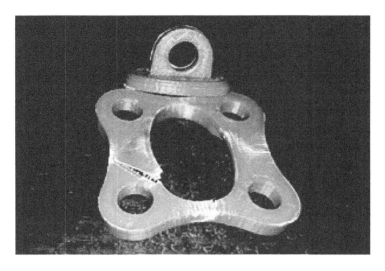

The intact compound lock and failed ring. Note that the ring failed in interlayer adhesion and tore along the extrusion lines. (Photo by author)

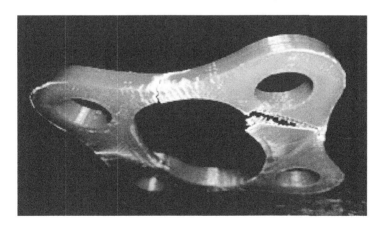

Closeup of the failed ring. Note that failures are primary in tension, and especially in interlayer adhesion. Stress lines can be seen near the failure points. (Photo by author)

The post-mortem examination highlights the role of interlayer adhesion failure as a primary weakness, and one that can be greatly influenced by the designer's choice of print orientation. Often, print orientation can be optimized by the breakdown of models into constituent parts so that each sub-part can be printed in an orientation that optimizes its strength.

Although it was not specifically part of the original idea for this experiment, it is also apparent that the one-piece eye also suffered from minor printability issues, and might easily have failed. At the least it may have required support material to print well on some printers, incurring penalties in print time, material usage and ultimately in print reliability.

In printing functional objects for utility, the role of the designer is difficult to overstate. When strength, printing reliability, print cost / speed, or weight are a factor, it is up to the designer to re-imagine the part in its ideal printable form and turn a simple shape into an object fully optimized for 3D printing and utility.

A 3d Printed tail-light mount for a dirt-bike, in PETG. Careful optimization of layer orientation and structural features allows for printing strong monolithic parts such as this angle bracket. It has been in rough service for 3 years as of publication and is still going strong. (photo by author)

APPENDIX IIC: INFILL AND STRENGTH

Designing functional objects to be 3D printed usually involves consideration of the required strength or durability of the finished object. Aside from the overall design and printing orientation, control or specification of the printing process itself is often a primary concern of the designer where strength will be an issue.

The printing process specification must address such questions as: What materials or processes can the part be printed in? How much or what type of infill is required? What are acceptable parameters for shell thickness, layer height and nozzle width? The need to address these and other factors to provide printing guidance to the end user leads us to the next question....just how is part strength influenced by printing or slicing parameters?

In this article, I will provide some very basic experimental results to examine from which we may draw some general conclusions. It is important to note that since the overall form and application of an object will be extremely influential in how infill and shell parameters affect its strength, these experiments serve only as guidelines and starting points. In critical objects, it is often necessary for designers to do their own testing to determine the acceptable range for printing parameters.

For this write-up, we will be using a standardized 6-mm cantilever beam, testing its load carrying capacity to the point of failure using a graduated water-load cylinder (white plastic bucket with a ruler in it). This crude load cell is sufficient to draw some broad conclusions from, but should not be taken as a substitute for application specific testing.

For the test, 4 load beams were prepared, each from the same model, but sliced with different settings. CURA was used as a slicer, as it gives good results and has a nifty 3D-preview feature to visualize the tool path. The parameters were adjusted for each beam to provide 3 beams with a 0.7-mm shell and 25%, 50%, and 75% rectilinear infill, as well as one beam with a 1.05-mm shell at 25% infill for comparison.

The load beams were fixed to the test platform and the variable load using steel wire, with the load hanging from the unsupported end of the cantilever beam, supported on the center fulcrum of the model. Loading was increased by adding water to the load cylinder, noting the level up to the point of failure.

Measurements were approximate, with a 1/8" (.125) estimated resolution. Failure modes were universally in tension, with failure beginning in the upper (stressed) skin, propagating as a tear downward, as expected.

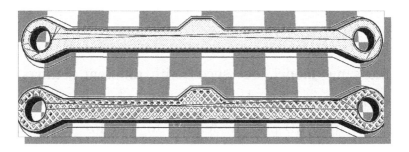

Test model in Cura slice view. Shown here with 2 perimeters, 25% and 75% infill versions. (Illustration by author)

Test setup and inset of using the calibrated measuring stick (Photo by author)

It is worth noting that although it was the weakest in load capacity before failure, the 25% -.7-mm beam deflected farther than any of the others before failure. This characteristic suggests that increased strength comes at the expense of flexibility in this type of printed structure.

The tests indicated an increase in strength as the infill was increased, with the greater gains between 25-50 percent. Gains between 50% and 75% were more modest, but still significant. Tests of the 1.05-mm shell 25% beam showed strength to be slightly less than the 50%.7-mm shell beam, at a slightly lower material usage.

The material used / strength relationship of the reinforced skin vs. increased infill tests were very nearly linear with this model. This was surprising, as I had anticipated that increasing the skin thickness would yield more strength per unit of plastic than increasing the infill.

The test results generally follow what empirical observation has shown, that infill density is a significant contributor to part strength, and that somewhere around 50%, a point of diminishing returns is reached. The precise optimal infill will vary based on part geometry, infill pattern, and the orientation of the load, but as infill rates increase it often becomes more profitable to scale or redesign the part instead of adding additional infill.

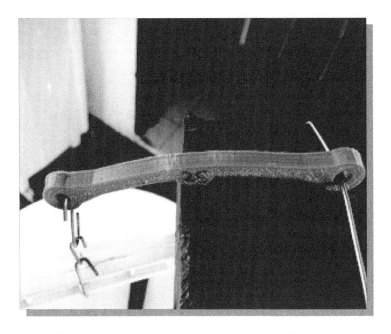

25% infill load beam under test. Note the stress lines forming on the upper surface, and significant bending. (Photo by author)

Infill Percent / shell thickness	Inches of water to failure	3-mm Filament used
25% / 0.7mm	3.125	.39m
50% / 0.7mm	4.25	.51m
75% / 0.7mm	4.75	.63m
25% / 1.05mm	4.0	.47m

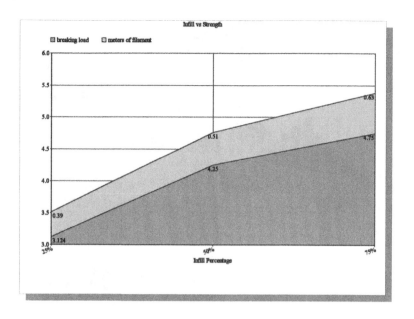

An interesting observation was that the lighter infilled parts were able to accommodate more bending prior to failure, even though they failed at lighter loads.

Among the many ways that the designer can influence the finished strength of an object, control of the printing process is the least direct. After all, an end user may take an STL file and slice it for any process, with widely divergent outcomes. For this reason it is important when designing objects for utility that the process and printing specifications should be detailed and provided with the model files as a description, read-me, or instruction file.

Break detail of the 25% infill beam with the 0.7-mm shell. Note the stress marks near the break. (Photo by author)

APPENDIX III: FASTENER TABLES

Hole sizes for threaded holes

Fastener or tap size (inch)	Hole size (inch)	Hole size (mm)	Fastener or tap size (mm)	Hole size (inch)	Hole size (mm)
0-80	0.047	1.191	3mm x 0.5	0.098	2.500
1-64	0.060	1.511	4mm x 0.7	0.134	3.400
2-56	0.070	1.778	5mm x 0.8	0.170	4.300
3-48	0.080	1.994	6mm x 1.0	0.205	5.200
4-40	0.090	2.261	7mm x 1.0	0.240	6.100
5-40	0.101	2.578	8mm x 1.25	0.272	6.900
6-32	0.110	2.794	8mm x 1.0	0.280	7.100
8-32	0.136	3.454	10mm x 1.5	0.343	8.700
10-24	0.150	3.797	10mm x 1.25	0.350	8.900
10-32	0.160	4.040	10mm x 1.0	0.358	9.100
12-24	0.177	4.496	12mm x 1.75	0.413	10.50
1/4-20	0.203	5.159	12mm x 1.5	0.421	10.70
1/4-28	0.220	5.556	14mm x 2.0	0.480	12.20
5/16-18	0.266	6.747	14mm x 1.5	0.500	12.70
5/16-24	0.272	6.910	16mm x 2.0	0.560	14.20
3/8-16	0.313	7.938	16mm x 1.5	0.580	14.70
3/8-24	0.328	8.334			
7/16-14	0.360	9.128			
7/16-20	0.391	9.922			
1/2-13	0.422	10.71			
1/2-20	0.453	11.51			
9/16-12	0.484	12.30			
9/16-18	0.516	13.09			
5/8-11	0.531	13.49			
5/8-18	0.578	14.68			
3/4-10	0.656	16.67			
3/4-16	0.688	17.46			

Hole sizes for US tapered screw sizes

For soft plastic use the next smaller hole size.

US Screw size (#)	Nominal screw diam- eter (inch)	Nominal screw diam- eter (mm)	Hard plastic Hole size (inch)	Hard plastic Hole size (mm)
0	0.060	1.5240	0.031	0.7875
1	0.073	1.8542	0.045	1.1430
2	0.086	2.1844	0.063	1.6002
3	0.100	2.5146	0.077	1.9558
4	0.112	2.8448	0.080	2.0320
5	0.125	3.1750	0.090	2.2860
6	0.138	3.5052	0.100	2.5400
7	0.151	3.8354	0.110	2.7940
8	0.164	4.1656	0.125	3.1750
9	0.177	4.4958	0.135	3.4290
10	0.200	4.8260	0.140	3.5560
11	0.203	5.1562	0.0145	0.3683
12	0.216	5.4864	0.155	3.9370
14	0.242	6.1468	0.170	4.3180
16	0.268	6.8072	0.190	4.8260
18	0.294	7.4676	0.220	5.5880
20	0.320	8.1280	0.235	5.9690
24	0.372	9.4488	0.266	6.7564

GLOSSARY:

Additive Manufacturing

Additive Manufacturing describes any manufacturing process where material is deposited in a controlled way to incrementally manufacture the end product. Some examples of additive manufacturing are: 3D printing, some types of welding, and thin film deposition. This type of process is characterized by minimal waste, since raw material is deposited only where needed. The reciprocal process, subtractive manufacturing, is represented in conventional machining and woodworking, sheet cutting processes, and other methods where material is removed from an object to achieve the desired form. Subtractive manufacturing is typically characterized by a high waste ratio, as removed materials are frequently rendered unsuitable for further use.

Adhesion (bed)

Bed adhesion refers to the degree that a printed part adheres to the bed surface, allowing printing to continue. Bed adhesion can be enhanced by using bed covering materials or adhesives that are compatible with the material being printed, a heated bed, or adhesion enhancing print features. The use of bed adhesion enhancers such as ABS-solvent slurry, hairsprays, or PVA glue can help to promote adhesion, though these are falling out of favor as PEI, BuildTak, and other specialty printing surfaces become more common.

Aliasing (errors)

In 3D printing, errors in the printing of a shape caused by the transition of one layer to the next. Very similar to pixelization or blockiness caused by resolution limitations in images. Aliasing errors in FFF/FFM/FDM printed objects are step-like artifacts (typically in the Z-axis) from the layers in a print, and can be minimized by using a minimal layer height.

Anisotropic

Non-uniform, different in some dimensions than others. Anisotropy in 3D printing usually refers to the directional strength characteristics of printed parts that result from being constructed in layers. 3D printed objects tend to suffer from a lack of tensile strength in the Z-axis.

Artifacts, Artifacting

Small errors, usually extra lumps or bulges of extruded material caused by imperfections in the printing process. Artifacts can be minimized by proper printer calibration, thoughtful design, and good slicing software settings.

Bridging

Printing over a gap between two structures, without any supporting structures underneath.

Delamination

Part failure occurring from separation of layers at the layer boundary, caused by inadequate design, excessive strain, or poor layer fusion. Generally this occurs along the Z-axis of FFM/FFF parts. See Grain and Anisotropic.

Extruder (drive)

The component of the printer that drives the plastic into the heater and nozzle. The extruder can be directly mounted to the nozzle assembly, or remotely mounted utilizing a drive tube. The remotely mounted systems are known as "Bowden" type extrusion systems.

Extruder (nozzle)

The part of the printer that applies plastic to the working surface. Also known as an effector, nozzle, or tool. Sometimes referred to as the "hot end".

FFF, FFM, FDM™ (additive manufacturing process)

Fused Filament Modeling / Fabrication or Fused Deposition Modeling™. The additive manufacturing process by which a filament is extruded and fused to build up an object being manufactured. This is the most common 3d printing modality.

G-Code

Machine specific action instructions, usually involving motion or temperature control in the FFM context. G-code is used in many kinds of robotic manufacturing, and may include machine specific commands for any controllable machine function. G-code often includes a command and a parameter... for example, 'G28 x50

Y50' could mean to move the x and y machine axes to 50 machine units (often millimeters) from their zero positions.

Glass transition temperature, (floor)

The (lowest) temperature at which a material begins to act as a fluid. See thermoforming.

Grain

Anisotropic strength characteristic of 3D printed parts caused by the Z-axis layerwise tensile weakness. Superficially similar to the "grain" of wood.

Gussets, Ribs, Bosses

Design features used to give additional strength to structures. A gusset is an intersecting structure, usually triangular in form that offers cross-span support to another plane by anchoring it diagonally to another intersecting plane. A rib is similar to a gusset, except it typically spans the entire space between two parallel planes or walls, reinforcing them both. A boss is a raised area, typically used to support a fastener attachment point. In plastic manufacturing it is often braced to the rest of the structure by ribs or gussets.

Hygroscopic

A Hygroscopic material is one with an affinity for moisture. Hygroscopic plastics must be kept dry or dried out prior to use in 3D printers, or steam bubbles will be formed in the extruder while printing. This produces in a bubbly or even foamy extrusion, often resulting in poor print quality.

Infill

Algorithmically generated "filler" printing that prints inside the shell of a model, adding strength and support for successive layers. Typically it forms a mostly empty "framework" structure.

Layer fusion

Adhesion by welding of one layer to the next in a 3D printed structure. Layer fusion is usually not quite as strong as the extrusion in other directions, giving rise to the anisotropic strength characteristics found in FFM printed objects.

Layer thickness (height)

The thickness that each extruded layer of a model is printed at. For example, a 1cm cube will consist of 50 printed layers if printed at a 0.2mm layer height. (See slicer, slice thickness)

Living Hinge

A simple hinge consisting of membrane spanning two parts to be hinged which flexes to accommodate the required motion.

Overextrusion

Surplus material deposited during the normal printing process. A slight surplus of extrusion from the rounded edge of the extrusion "bead" resulting in a small increase in size of the part at its edges. As a general rule of thumb, the model should be undersized by one fourth to three fourths of the nozzle width setting in the X-Y plane.

Post-Print Processing

Sanding, support material removal, assembly, or other processes required after printing to complete the manufacturing of a printed object.

Pre-Print Processing, Pre-Processing (slicing)
Conversion of a 3D model into specific instructions (typically G-code) given to a manufacturing device such as a 3D printer to cause it to perform the actions necessary to manufacture the specified object.

Software / Production stack (software)
A software system covering the end to end processing of data. In 3D printing, it could be an integrated system of design, slicing and printing software. Not to be confused with the colloquial programming use of stack (as in "stack overflow"), which refers to a type of transient storage for data being processed.

Refactor, refactoring, factoring
Dividing a model into individual parts printed separately, to be assembled after printing is complete. This is usually done to allow printing each sub-component in its ideal orientation for printability, strength, and printer size limitations. A primary paradigm for functional design, it is especially useful for optimizing strength and minimizing printing time.

Shell, Perimeters, Solid layers

The solid skin portion of a 3D printed object. In the X-Y plane, this is the number of solid perimeters in each layer. In the Z-axis, this is the number or thickness of solid layers at the upper and lower surfaces of a printed piece.

Slicer

A piece of software that processes a 3D model into a series of slices ascending along the Z-axis, and creates detailed instructions for operating the printing nozzle for each slice.

Slice thickness

The Z dimensional thickness of the slices that a 3D model is decomposed into by the slicing software (see slicer, layer thickness)

Support Material

Printed structures that are not intended to be part of the finished model, printed only to support other structures. Support structures will typically be manually removed or dissolved away in a solvent bath after printing. Support structures may be added by the designer, or may be algorithmically generated by the slicing software. The minimization of support material requirements is often a primary design goal in order to improve printing speed, reliability, and to reduce material use.

Thermoforming

Molding, shaping, or forming a material at or above its glass transition temperature. When cooled, the material will retain its new form. Some printed plastics such as PLA can be easily thermoformed after printing by briefly immersing them in boiling water and forming them into the desired shape.

Warping (bed adhesion)

The tendency of some plastics, notably nylon and ABS, to warp and peel away from the bed during printing, causing distortion or print failure. Warping can be mitigated by design, print process changes, heated beds and enclosures, and by the use of adhesion enhancing features such as brims or rafts.

ABOUT THE AUTHOR:

An early adopter of additive manufacturing and an innovator in the 3D printing space, I divide my time between my home in Alaska and the Caribbean. My favorite authors include Neal Stephenson, Dewey Lambdin, George MacDonald Fraser, William Gibson and Tom Robbins, to name a few. I am interested in pursuing collaborative projects in additive manufacturing, pervasive computing, and many other technology related fields.

Feel free to contact me at the address below regarding consulting or interesting collaborative projects.

cliff@functionaldesignbook.com

Also, See my printable flying planes at threedsy.com, or check out some cool flying freebies at thingiverse.com/exosequitur

HAAARRRRR!

If you received this book through file sharing and found it worthwhile, please go outside and do something unusually kind for someone who looks like they could use it. Barring that, consider supporting my creative endeavors by buying some of my designs at threedsy.com, by tipping me some coffee or buying me lunch. A good review on your favorite bookseller website is also a worthwhile kindness.

BTC

DOGE

BTC: 1HcADp5yRNNBVLkwGpeo9TAf4yRnUVxsTW
DOGE: DSxf3VfbsLMr7yqAjqFiXN8LaoMSHLu18p

The local constabulary and I thank you in advance for helping to keep me from the streets, where I would otherwise likely be found making a general nuisance of myself.

READERS NOTES

~∞~

Made in the USA
Las Vegas, NV
30 September 2024

96032106R00134